Malaysia, Modernity and the Multimedia Super Corridor

Based on fieldwork in Malaysia, this book provides a critical examination of the socio-spatial transformation of the country's main urban region. In the late twentieth century Kuala Lumpur experienced a spectacular redevelopment of urban space while the city formed part of an extended metropolitan area. Particular attention is paid here to the development of the Multimedia Super Corridor, a high-tech zone which extends fifty kilometres from the capital, and includes two new 'intelligent' cities, Putrajaya and Cyberjaya.

This study first provides a theoretical reworking of geographies of modernity and details the emergence of a globally-oriented, 'high-tech' stage of national development. The Multimedia Super Corridor is framed in terms of a political vision of a 'fully developed', modern Malaysia before the author traces an imagined route through surrounding landscapes in the late 1990s. Beginning at the site of the world's tallest building, the Petronas Twin Towers, the reader is then taken southward to examine Putrajaya and Cyberjaya as spaces of government for 'intelligent' citizens. The journey ends at the new international airport where goals of 'global' connectivity legitimised the destruction of indigenous territory.

As the first book-length academic analysis of the development of Kuala Lumpur and the construction of the Multimedia Super Corridor, this work offers a situated, contextual account which will appeal to all those with research interests in Asian Urban Studies and Asian Sociology.

Tim Bunnell is Assistant Professor in the Department of Geography at the National University of Singapore.

Routledge Pacific rim geographies
Series editors: John Connell, Lily Kong and John Lea

Malaysia, Modernity and the Multimedia Super Corridor

A critical geography of intelligent landscapes

Tim Bunnell

Routledge
Taylor & Francis Group

LONDON AND NEW YORK

First published 2004
by Routledge
2 Park Square, Milton Park, Abingdon, Oxon, OX14 4RN

Simultaneously published in the USA and Canada
by Routledge
270 Madison Ave, New York, NY 10016

Routledge is an imprint of the Taylor & Francis Group

Transferred to Digital Printing 2006

Typeset in Sabon by LaserScript Ltd, Mitcham, Surrey

British Library Cataloguing in Publication Data
A catalogue record for this book is available from the British Library

Library of Congress Cataloging in Publication Data
Bunnell, Tim
   Malaysia, modernity, and the multimedia super corridor : a critical
geography of intelligent landscapes / Tim Bunnell.
      p. cm.
Includes biographical references and index.
   1. Regional planning–Malaysia. 2. Information technology–Malaysia.
3. Malaysia–Economic policy. 4. Malaysia–Economic conditions. I.
Title.
   HT395.M4B86 2004
   307.1′216′09595–dc22

                                                              2003015891

ISBN 0–415–25634–8
Printed and bound by CPI Antony Rowe, Eastbourne

For Craig and Megan Bunnell

Contents

Figures

Acknowledgements

I have been fortunate to receive help and support from many people at each step along the protracted route to completing this book. In Nottingham, where it 'all started' in 1995, I remain grateful to Chris Abel, Louise Crewe, Denis Linehan, David Phillips, Adam Swain, Charles Watkins and two generations of 'A28ers'. I have even bigger academic debts in the School of Geography, however – to Dave Matless and Steve Daniels – whose wonderful combined supervision I now appreciate all the more as I attempt to advise students myself. The way forward from doctoral research and writing was in part illuminated by helpful comments from two anonymous referees and the examiner's recommendations of John Allen.

I am grateful to the Economic Planning Unit of the Prime Minister's Department of Malaysia for eventually granting me permission to carry out fieldwork in 1997. My time in and around Kuala Lumpur would have been much less productive in the absence of advice from Lee Boon Thong, Jamilah Mohamad, Tan Wan Hin and other friends at Universiti Malaya's Department of Geography. Financial assistance from the Dudley Stamp Memorial Fund for fieldwork in Malaysia is gratefully acknowledged. The 'field' has been an even more inviting place more recently thanks to Bram, Emily and not-so-little Arthur Tan. Much more difficult is a long overdue expression of appreciation to Huey Yap for lots of stuff for which I never could find the words ...

The National University of Singapore has been (and remains) a wonderful place to geo-graph. Those colleagues, past and present, who have helped to shape what follows in this book – but who, of course, are not to blame for it – include Paul Barter, T. C. Chang, Neil Coe, Kris Olds, James Sidaway, Brenda Yeoh and Henry Yeung. NUS research projects have provided me with the benefit of collaborative insights from outside 'Geography' (Andrew Hardy and S. M. A. K. Fakhri) and even from beyond Singapore (Morshidi Sirat). In a rather different collaboration, Alice Nah is my sternest but favourite critic. Phil Kelly and Lily Kong kindly encouraged me to work towards publication, but I am equally thankful to Lisa Law and Victor Savage for giving me the confidence needed for completion.

The following have kindly given permission to reproduce material: Figure 1.1 Multimedia Development Corporation; Figure 4.2 Malaysia Airlines; Figure 4.3 © *The Guardian*; Figure 4.4 MEC Marketing Sdn Bhd; Figure 5.2 Putrajaya Holdings Sdn Bhd; Figures 5.3 and 6.1 *The Star*, Malaysia; and excerpts from a poem in *Utusan Malaysia*, Consumer's Association of Penang. Lee Li Kheng's cartographic work is also greatly appreciated. Lim Kean Fan and Josephine Then provided timely assistance with referencing. Thanks, finally, to the patient people at Routledge for putting up with all the missed deadlines.

Abbreviations

API	Air Pollution Index
ASEAN	Association of Southeast Asian Nations
BA	*Barisan Alternatif* ('Alternative Front')
BN	*Barisan Nasional* ('National Front')
CDC	Community Development Centre
COAC	Centre for Orang Asli Concerns
DAP	Democratic Action Party
EIA	Environmental Impact Assessment
EOI	Export-Oriented Industrialisation
EPZ	Export Processing Zone
ERL	Express Rail Link
FELDA	Federal Land Development Authority
FMS	Federated Malay States
GTA	Golden Triangle Area
HICOM	Heavy Industries Corporation of Malaysia
IAP	International Advisory Panel
ICT	Information and Communications Technology
ISA	Internal Security Act
ISI	Import-Substituting Industrialisation
ISIS	Institute of Strategic and International Studies
IT	Information Technology
JHEOA	*Jabatan Hal Ehwal Orang Asli* ('Department of Orang Asli Affairs')
KLCC	Kuala Lumpur City Centre
KLIA	Kuala Lumpur International Airport
KLLC	Kuala Lumpur Linear City
KLMA	Kuala Lumpur Metropolitan Area
KMM	*Kesatuan Melayu Muda* ('Union of Malay Youth')
LEP	Look East Policy
MAMPU	Malaysian Modernisation and Management Planning Unit
MCA	Malaysian Chinese Association
MCP	Malayan Communist Party
MDC	Multimedia Development Corporation

MIC	Malaysian Indian Congress
MIMOS	Malaysian Institute of Microelectronic Systems
MSC	Multimedia Super Corridor
MTDC	Malaysian Technology Development Corporation
NDP	National Development Policy
NEP	New Economic Policy
NERP	National Economic Recovery Plan
NGO	Non-governmental organisation
NIC	Newly Industrialising Country
NIDL	New International Division of Labour
NITA	National Information Technology Agenda
NITC	National Information Technology Council
NLD	National Landscape Department
NOC	National Operations Council
NST	New Straits Times (and New Sunday Times)
NTT	Nippon Telegraph and Telephone Corporation
NUPW	National Union of Plantation Workers
PAP	People's Action Party
PAS	*Parti Islam SeMalaysia* ('Pan-Malaysia Islamic Party')
PERNAS	*Perbadanan Nasional Berhad* ('National Corporation')
PKMM	*Partai Kebangsaan Melayu* ('Malay Nationalist Party')
PNB	*Permodalan Nasional Berhad* ('National Equity Corporation')
POASM	*Persatuan Orang Asli Semenanjung Malaysia* ('Peninsula Malaysia Association of Orang Asli')
PPP	People's Progressive Party
RIDA	Rural and Industrial Development Authority
RM	Ringgit Malaysia
SOE	State Owned Enterprise
STC	Selangor Turf Club
TNC	Transnational Corporation
TPD	Total Planning Doctrine
UMNO	United Malays National Organisation
UKM	Universiti Kebangsaan Malaysia
UPM	Universiti Putra Malaysia (formerly Universiti Pertanian Malaysia)
USM	Universiti Sains Malaysia
UTM	Universiti Teknologi Malaysia

1 Introduction

On 1 August 1996, Malaysian Prime Minister, Dato' Sri Dr Mahathir Mohamad announced that a 50 km corridor of land extending southwards from the federal capital, Kuala Lumpur, had been designated as a special zone for the development of information and multimedia technology (Mahathir, 1996a). The Multimedia Super Corridor (MSC) gave a high-tech urgency and apparent coherence to existing processes of social and spatial development in Malaysia's main metropolitan region. MSC was delimited by two existing high-profile megaprojects. At the northern end was Kuala Lumpur City Centre (KLCC), a 'city-within-a-city' commercial development which included the world's tallest building, the Petronas Twin Towers. At the other end, MSC's southern-most point was Kuala Lumpur International Airport (KLIA) which eventually opened in 1998. In between these poles, work was under way on a new 4,581 hectare federal government administrative centre known as Putrajaya (see Figure 1.1). At the launch of MSC, Mahathir announced that Putrajaya would be accompanied by an 'intelligent city' for high-tech companies. Cyberjaya, as this became known, and Putrajaya were to be connected up to either end of the new urban corridor. This book provides a critical geographical analysis of the intelligent landscapes of the Multimedia Super Corridor, a project that Mahathir proudly announced as a 'world first' (Mahathir, 1996a).

For all its supposed uniqueness, the MSC project was one of a succession of attempts around the world to create a Silicon Valley-style 'technopole' (see Castells and Hall, 1994; Winner, 1992). *Wired* magazine identified and evaluated some forty-six would-be 'venture capitals' in mid-2000 (Hillner, 2000). These otherwise diverse locales share two key characteristics. First, is the attempt to attract and foster innovative economic activity. They seek to become 'hubs' for regional operations of existing leading-edge high-tech companies and breeding grounds for new ones. Second, is a logic of connectivity, a perceived necessity of 'plugging in' to a global 'space of flows' (see Castells, 1996). MSC was built upon a 2.5–10-gigabit digital optical fibre backbone enabling direct high-capacity links to Japan, the USA and Europe (Multimedia Development Corporation, 1996a). The particular vogue for innovative and wired urban spaces in the Asian region in the

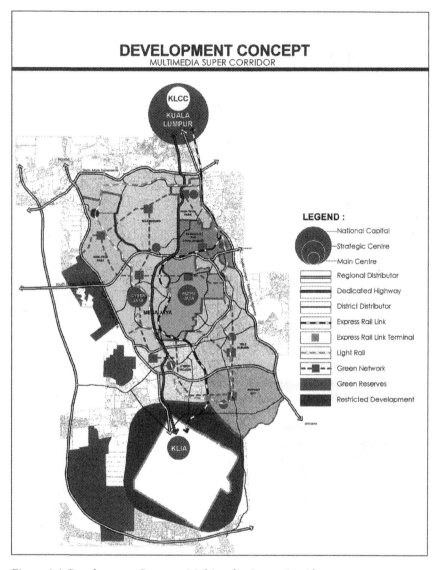

Figure 1.1 Development Concept: Multimedia Super Corridor.

Source: Reproduced with permission from Multimedia Development Corporation.

mid-1990s led two commentators to refer to a generalised 'Siliconisation of Asia' (Jessop and Sum, 2000).

Malaysia's high-profile Siliconisation in and through MSC drew upon celebratory accounts of a supposedly new era based on information and communications technology (see, for example, Gates, 1995). Political speeches, marketing brochures and exhibitions presented an information society which would not only be more affluent, but also variously simpler, cleaner, more environmentally-friendly, more efficient and more egalitarian. The ability to overcome constraints of space and time, a familiar theme in utopian imaginings of technological change, was projected in domains ranging from health care provision (through 'telemedicine') to public sector services (using 'electronic government'). Specifically electronic promises were articulated in new vocabularies of technological optimism speaking, for example, of 'smart schools' and a variety of 'intelligent features'. As 'a global facilitator of the Information Age' (Mahathir, 1996a), MSC would lead Malaysia to a national 'Multimedia Utopia' (Multimedia Development Corporation, 1997a: 4).

Despite this utopianism, the very ideal of connecting up to an Information Age whose imagined centre lay elsewhere appeared to reaffirm Malaysia's global peripherality. Technological utopian futures were, of course, premised on advancing from a not-so-idealised present. More than this, any contemporary technological lack was to be addressed in MSC by importing 'progress'. The state-controlled Malaysian press in 1997 charted the number of companies applying for 'MSC status', celebrating each 'world class' foreign addition as a step closer to the indigenous nurturing of innovative and creative selves – what have been referred to elsewhere as '*Homo Silicon Valleycus*' (Thrift, 2000a: 688). Some of this breed sat on MSC's International Advisory Panel, legitimising the corridor and steering it towards its 'intelligent' evolutionary apex.[1] The centre of older 'biological geographies' (see Rose, 1999: 38) had thus shifted from Western Europe to Southern California. Yet the locus of an imagined Silicon modernity of global flows and connectivity remained very much 'out West'.

In this book, I argue that MSC cannot be adequately understood as either an expression of a paradigmatic global shift to a new techno-economic era or in terms of the expansion of a modern 'West' into a 'non-Western' periphery. In the first place, it is important to reject any suggestion of a wholesale transition to a distinct new technological epoch. As Nigel Thrift (1996a) has reminded us, writing about new technology has a propensity towards notions of rupture and 'revolutionary' understandings of social change. More nuanced accounts attend to the messy complexity of social and political constructions of technology in particular places rather than 'explaining' transformation in terms of the emergence of a generalised Information Age. Second, reinscribing a generalised West–other cartographic dichotomy makes little sense in a world where social, economic and technological processes are increasingly global in scope. 'Western' knowledge, as Wendy Mee has argued, is always already part of Malaysia's 'in here' (Mee, 2002: 70). In addition, to reduce transformation to the effects of some supposedly external, modern(ising) force is to deny the agency of in situ individuals and institutions. A critical analysis of Malaysia's Siliconising landscapes, in other words, cannot proceed in the absence of the *human* geographies of 'real' places.

The focus of this research, then, is the people, places and landscapes of Malaysia's Multimedia Super Corridor. This is not to deny the significance of broader advances in information and communication technologies nor connections with distant locales such as in the USA. Clearly, emergent informational economies rationalise investment in Silicon Valley-style technopoles; and the form that these take in Malaysia, as elsewhere, is shaped in part by flows of knowledge, capital and people from other(s') places. However, I seek to combine these insights with a more grounded perspective. This means not so much seeing MSC as diagnostic of somehow more fundamental or more important processes and forces. Rather, it means attending to how 'intelligent' landscapes and lives in MSC are shaped by people and processes at a variety of scales. It also means considering how ostensibly 'local' MSC norms and forms are bound up with networks of culture and power, in reciprocal – if asymmetrical – relation to elsewhere. Such an approach demands a conceptualisation of locality which is not geographically-bounded. Work in human geography has shown how place (Massey, 1993), landscape (Schein, 1997) and the local (Mitchell, 2001) can be spatialised rather differently – in terms of an 'articulated moment in networks that stretch across space' (Schein, 1997: 662). Following such conceptual qualification, this work may be understood as a local study of MSC landscapes and as an analysis of the modernity of a particular place and time.

Reworking modernity

To speak of modernity in terms of spatial and temporal specificity may conventionally be considered as somewhat contradictory. 'Modernity' has been an occasion for the very kinds of portentous theorisation and epochal claims from which I have sought to distance the approach adopted in this book. However, recent historical work in geography has demonstrated how a consideration of 'spaces of modernity' sidesteps some of the pitfalls of 'grand' theory (Ogborn, 1998: 2; see also Nash, 2000). As Miles Ogborn has put it, situating the concept involves 'fracturing modernity as a totality by contextualising it in terms of specific histories and geographies' (Ogborn, 1998: 38). Modernity, in this way, is not claimed as a generalised new period or stage, but as a way of framing specific transformation; it is less an attempt to formulate universally-applicable explanation, than a perspective for the study of particular geo-historical spaces.

There are three further geographical dimensions to the critical reworking of modernity in this book. The first, and most straightforward, is the study of modernity beyond the metropolitan centres of 'the West'. As Nash and Graham (2000: 2) have pointed out, 'While there is increasingly sensitivity to social relations based on gender, race and class, less attention has been paid to the different historical geographies of modernity beyond the metropolis, in the margins of Europe or in the non-European world'.

I consider the modernity of the high-tech spaces of 1990s Malaysia. This implies more than just a remapping of modernity into territories of the 'non-European world'. Rather – and this is the second geographical dimension – I am concerned with the active role of space in the realisation of new cultural-economic processes and practices. MSC landscapes are considered to form part of attempts to foster new ways of being and seeing. Modernity here, in other words, is understood as experienced *through* rather than merely *in* space. This, in turn, leads on to a third dimension: differentiated experiences of modernity in different times and places. It is important to clarify that I am not positing the existence of bounded, alternative (or 'non-Western') modernities. Rather, I argue, a multiplicity of experienced modernities is forged through shifting networked relations of interconnection which resist bounded topological packaging – and which therefore unsettle the very binary of West–the rest.

The plurality implied by taking seriously geographical as well as historical specificity might appear better suited to the frame of 'postmodernity' than to the perspective or stage that this is implied to have moved beyond (see Harvey, 1989). For Joel S. Kahn (1998: 83), however, a combination of accelerated capitalist transformation, rapid bureaucratic rationalisation and political commitment to discourses of modernisation in Malaysia in the 1990s signalled a 'reworking of the modern project' that was more accurately labelled 'neo-' than 'post-modern'. On the one hand, Malaysia was politically committed to joining the 'fully developed' ranks of mostly Western countries (Mahathir, 1993). On the other hand, there were national- and regional-scale reasons for political reworking of Euro-American-centred meanings of 'development' and 'modernisation'. Economic depression in the mid-1980s had been followed by a period of unprecedented growth in Malaysia.[2] Not only did this appear to vindicate Mahathir's economic policies; it also dampened opposition to the increased authoritarianism which accompanied his administration in the late 1980s. The year 1995 saw an electoral landslide for the ruling *Barisan Nasional* ('National Front') coalition (Liak, 1996). This represented a personal triumph for Mahathir who, like the country as a whole, appeared finally to have emerged from a period of political insecurity and uncertainty (Harper, 1996). A series of monumental state development projects, including the MSC, were founded not only on this new stable personal and national ground, but also on a regional belief that history and geography themselves were on Malaysia's side. The twenty-first century, after all, would be the 'Asian Century' (*New Perspectives Quarterly*, 1997).

The economic performance of Asian 'tigers' and 'dragons' emboldened regional elites to make truth claims about the leading edge of global change in the 1990s. Stories of (East) Asian 'miracles' (World Bank, 1993) fuelled re-imagination of the mystical, sleepy (post-)colonial Orient as, at once, a new threat to Euro-American supremacy and a new paradise of economic opportunity (see Dirlik, 1998). Thus, while there was some critique of the human rights and environmental records of rapidly-developing Asian economies (Seabrook, 1996), there was also no shortage of 'pilgrims' eager to learn the 'Asian way' (Webb, 1996). One of the most vociferous advocates of 'Asian values', Prime Minister Mahathir was a key actor in the discursive as well as material construction of an ordered and disciplined 'East' in opposition to a 'West' imagined as decadent and decaying (Mahathir, 1996b; see also Mahathir and Ishihara, 1995). Rather than the post-Soviet Union 'end of history' (Fukuyama, 1992) and a liberal democratic 'capture' of modernity for itself (Held, 1992), Asian values or 'Renaissance' (Anwar, 1996) implied the return of history and the possibility of alternative modernities. Imaginatively and cartographically realigned as part of a 'new Asia' (Noordin, 1996: 6), Malaysia and other countries in the region could now plot courses to their own (neo-)modern futures.

This is not to suggest that modernity's multiplicity is either reducible to elite discursive reworking or specifically territorialised at the national scale. First, Asian values and other discourses associated with political elite groups merely form part of broader shifting constellations of political and cultural forces (Kahn, 1997). There is clearly a connection between political discourse and practices of the 'national self' through processes of 'cultural subjectification' (Holden, cited in Wee, 2002: 19). In this book, I give attention to the 'cultural' dimensions of (self-)regulation and seek to contribute to recent work subjecting discursive practices to post-structuralist theoretical scrutiny (see Yao, 2001a: 3). Yet it is also important to consider that elite discourses themselves emerge from the ongoing reworking of 'development'. A range of political and cultural authorities including – but certainly not only – state actors contest appropriate aims of and means to social transformation, including the role of the state. Aihwa Ong's recent work is instructive here in giving attention to specific 'modes of biopolitical regimes', particular relations of state, culture and capitalism in Southeast Asia (Ong, 1999: 35). If modern 'bio-power' is oriented to the enhancement of the lives and welfare of national populations (Foucault, 1990: 143), Ong considers particular forms and mixes of political (non-)intervention in the government of populations and relations with global capital (Ong, 1999). In this way, it is possible to situate elite discourses in broader governmental rationalities – appropriate aims of and means to development – in particular times and places.

However, second, the diversity of bio-political regimes defies simple mapping onto nation-state boundaries. It is increasingly possible to identify differentiated spaces or zones of bio-political regimes within and across national boundaries, thus fragmenting the nation-state as a coherent 'territory of government' (Rose, 1996a). Aihwa Ong (1999: 21) elaborates this process in relation to the emergence of what she terms the 'postdevelopmental state': people and places within national territories are subjected to variegated degrees of state and corporate regulation as well as to different modes of state power. The effect of postdevelopmental strategies may therefore be to foment new patterns of exclusion and inequality – there is clearly nothing inherently celebratory about the reworking of modernity beyond 'the West'. In addition, as we will see, political tensions arise from post-developmental disjunctures between ostensibly *national* development and bio-politically privileged trans- and sub-national spaces.

These issues motivate a critical geographical engagement with the development of the Multimedia Super Corridor. Work in anthropology, urban planning and geography has already yielded some social scientific analysis of the MSC project (Hutnyk, 1999; Mee, 2002; Boey, 2002; Bunnell, 2002a and 2002b). However, existing book-length treatments have been at best descriptive and/or prescriptive if not explicitly promotional (Ibrahim and Goh, 1998; Norsaidatual *et al.*, 1999). One, carrying dual subtitles of 'what the MSC is all about' and 'how it benefits Malaysians and the rest of the world' lauds MSC as an 'engine of growth for Malaysia's Information Technology (IT) industry' and even provides details of how companies can apply for 'MSC status' (Ibrahim and Goh, 1998: 130). In *Malaysia, Modernity and the Multimedia Super Corridor* I do not so much contest boosterist texts, as seek to incorporate them in diagnosing prevailing rationalities of development. I do not attempt to evaluate MSC in its own terms – to draw up, for example, a balance-sheet of its planned versus actual completion schedules, investment targets or innovation achievements – nor do I explicitly propose an alternative 'way forward'. My aim is to provide a critical examination of both the processes and systems of evaluation which underlie the emergence of MSC landscapes and their socio-spatial outcomes.

Landscape is a key concept in this critical cultural geography of MSC, denoting both material space and governmental strategies to induce new ways of seeing and being. On the one hand, MSC is a roughly 50 by 15 km *site* with new urban forms and intelligent infrastructure to facilitate new forms of living and working. On the other hand, MSC landscapes in various media make known ideals of conduct for the (self-)realisation of suitably high-tech subject-citizens. Particularly during the main period of fieldwork in 1997 upon which this book is based, the *sights* of MSC – in brochures, advertisements and public exhibitions – visualised ideals of individual and national conduct. MSC's intelligent landscaping, then, may be understood normatively rather than in purely technological or infrastructural terms. Both within and beyond the corridor, individuals and groups varied – and continue to vary – in their ability (and willingness) to realise themselves in 'intelligent' ways. If a reworking of modernity is understood in terms of the constitutive extra-local connections of MSC landscapes, therefore, this critical geography also brings into view emergent social and spatial dividing practices at a variety of scales.

The structure of the book

The book is divided into two parts. Part I provides an extended theoretical and geo-historical framing of the Multimedia Super Corridor. 'Framing Malaysia: concepts and context' consists of two chapters. Chapter 2 considers how the concept of modernity in social science has been taken to refer to – and has, in turn, constructed – a spatial and temporal division between the modern and its 'others'. I show how modernity can be productively applied to analyses beyond 'the West' without reinscribing a geographical opposition of Western and 'other' modernities. Drawing upon recent work in geography and anthropology, I elaborate how modernity's multiplicity may be conceptualised in terms of dynamic, relationally-constructed rationalities of government. These make known appropriate forms of modern (bio-)power at the level of the population involving the state and a range of other authorities. 'Global' technological and economic transformations which are commonly understood to have effected a diminution of state power are shown, rather, to have fragmented the nation-state as a unified territory of government. MSC is thus framed in terms of a post-developmental segmentation into variegated zones of bio-political investment. The chapter concludes with an elaboration of how the concept of landscape in cultural geography may be deployed to analyse the multi-scalar governmental role of MSC and associated socio-spatial dividing practices.

Chapter 3 is contextual, sketching the post-colonial transformation of Malaysia and, ultimately, the processes and politics which rationalised investment in a Silicon Valley-style intelligent urban space in the last decade of the twentieth century. The chapter initially focuses upon the formation of a socially- and spatially-diverse nation-state in which the majority 'Malay' community came to occupy a 'special political position' on account of claims to territorial indigeneity. The early 1970s saw the political crystallisation of distinct new aims of and means to government. In the wake of ethnic riots in 1969, the New Economic Policy (NEP) initiated a state-led programme of wealth redistribution to promote more ethnically and regionally 'balanced' national development. Dr Mahathir, came to power in 1981, at the mid-point of NEP's planned twenty year course. Mahathir's administration has seen the expiry of NEP, its replacement with the National Development Policy (NDP) and the articulation of a new long-term goal of making Malaysia into a 'fully developed country' by the year 2020 (Mahathir, 1993). As a federal government-led 'test-bed' for the use and development of information technology, MSC imaginatively connected contemporary Malaysia with an ideal(ised) multicultural society of 'Vision 2020'. Yet, rather than reducing MSC to the aspirations of a 'visionary' Prime Minister (or to a unified state logic), I am concerned with the discourses and conceptualisations that constructed MSC as a valid means of national development. MSC is thus contextualised as part of a broader geo-historical positioning of 1990s Malaysia.

Part II of the book, 'On Route 2020', focuses attention down to the material spaces of MSC development. It consists of three chapters each centred upon a particular site (or sites). The intention is that these chapters can be read either individually – as separate case studies – or together, as part of a larger national development project. 'Route 2020' was the nick-name given by planners in 1997 to a 'Dedicated Highway' which was proposed to link Kuala Lumpur with the new Kuala Lumpur International Airport (KLIA) at the opposite end of MSC (see Figure 1.1). In the political context of 1990s Malaysia, I understand Route 2020 as both a path to a state-led vision of Malaysian modernity and also in terms of physical connectivity in the 50 km corridor. Together, Chapters 4 to 6 can be read as a southward journey along this imaginative transect.

The route begins in Chapter 4 at what became the northern-most point of the MSC and one of the 'monuments' of the Mahathir era. Kuala Lumpur City Centre (KLCC), an urban megaproject closely associated with Prime Minister Mahathir, included the world's tallest building, the Petronas Twin Towers. I combine iconographic interpretation of the building with post-structuralist understandings of its geo-historical effects. The Petronas Towers performed a symbolic civic role both internationally and domestically: marking the city and nation on global maps as a modern and 'investible' metropolis; and demonstrating to citizens that *Malaysia boleh*, a 'can do' attitude free from the supposed shackles of (neo-)colonial inferiority. However, I also consider the ways in which intended meanings of the building were reworked and symbolically contested during its construction. The rise of the Petronas Towers on the skyline of the capital city elevated the building site/sight to a centrality in critical views on urban and national development more broadly. Nonetheless, KLCC and the Petronas Towers, in particular, I suggest, played an active role in the 'global' reorientation of Malaysia(ns) in the 1990s. As part of a normative landscape, the Petronas Towers were bound up with the construction of 'world class' ways of being and seeing.

From Kuala Lumpur City Centre (KLCC), in Chapter 5, I move southwards to the site(s) of the new MSC cities, Cyberjaya and Putrajaya (see Figure 1.1). While conceived to perform very different functions, plans for and representations of the cities together provide important insights about rationalities of MSC development. I detail the infrastructural and (self-)regulatory characteristics of these so-called 'intelligent cities' and their intended socio-cultural subject outcomes. 'Intelligent citizens' here, I argue, were imagined as those not merely adept in the use of existing information and multimedia technology, but also as capable of innovation and continual (self-)learning. Yet I also demonstrate that, in mid-1990s Malaysia, planning came to extend beyond the provision of environments conducive for technological development. Official representations of the MSC as a pioneer 'test-bed' also performed an important political function. On the one hand, what were cast as potentially harmful social effects of the 'global'

MSC experiment could be contained; on the other hand, official discourses suggested that 'successful' Malaysian constructions of technology could (and would) be extended across the national territory. Media performances of MSC as a national landscape served to obscure the scalar and spatial tensions of a privileged zone of bio-political investment.

Chapter 6 tests these technological utopian imaginings in the rapidly-transforming field of the MSC in the late 1990s. I highlight socio-spatial divisions associated with the emergence of 'intelligent' MSC landscapes and authoritative rationalities of development more generally. Building upon 'indigenous' critical reflection on urban development, I focus upon Indian plantation workers who were resettled to make way for Putrajaya and – moving progressively further south along Route 2020 – First Peoples groups impacted by the construction of a new highway and the Kuala Lumpur International Airport (KLIA). Landscapes of modernity are thus shown to be bound up with new geographies of exclusion. Two interrelated socio-spatial dividing practices run through the chapter: first, between those people who are enticed to putatively utopian urban environments and those who are displaced and/or excluded from such privatised spaces; and, second, between those people (and places) who are deemed capable of realising themselves in 'intelligent' ways versus those who are unable or unwilling to do so. There thus emerges a complex moral geography of graduated citizenship associated with modernities of MSC-Malaysia.

Each of the three chapters in Part II has been revised substantially in the light of new material on the sites and sights of MSC since the main period of fieldwork in 1997. This does not represent an attempt to bring coverage of these spaces 'up to date'. Rather, I have sought to use primary and secondary data collected during 1999–2001 to extend analysis of each geo-historical site/sight. Significant economic and political transformation can now be traced to events beginning to unfold during fieldwork in 1997 (see Hilley, 2001). I have not sought to provide a sustained treatment 'On Route 2020' of either the economic 'crisis' or the political contest which led to the sacking of Deputy Prime Minister Anwar Ibrahim and formation of a new opposition political coalition. If these events mark a key turning point in the development of contemporary Malaysia, then this book is perhaps best understood as an historical cultural geography of the modernity of pre-crisis MSC.

However, in keeping with the conception of social transformation outlined here – and without downplaying the political (or political economic) importance of the 'Anwar Affair' or *Reformasi* – I am wary of pronouncements of radical geo-historical change. In the conclusion to the book, I do sketch post-1997 financial and political developments and their relation to MSC. However, I suggest that a non state-centred approach and the expanded conception of 'the political' which this implies allows us to re-view the events in terms of the ongoing contested reworking of rationalities of government. To speak of modernity, of course, is to speak of transformation itself (Berman, 1982). Perhaps fear of intellectual obsolescence goes some way to explaining why so much work on modernity has focused on grand scale trends and forces. Yet I insist that it is only by attending to the small spaces that we begin to understand the rationalities of development and cultural politics of multiple experienced modernities.

Part I

Framing Malaysia

Concepts and context

2   Modernity, space and the government of landscape

One of the most problematic legacies of grand Western social science ... is that it has steadily reinforced the sense of some single moment – call it the modern moment – that by its appearance creates a dramatic and unprecedented break between past and present. Reincarnated as the break between tradition and modernity and typologised as the difference between ostensibly traditional and modern societies, this view has been shown repeatedly to distort the meanings of change and the politics of pastness. Yet the world in which we live – in which modernity is decisively at large, irregularly self-conscious, and unevenly experienced – surely does involve a general break with all sorts of pasts.

Arjun Appadurai (1996: 2–3)

When I was a child growing up in a predominantly Chinese city in Malaysia, it seemed as though we were always trying to catch up with the West, represented first by Britain and later by the United States. Although Malaya gained independence from the British in 1957 (and became Malaysia in 1962), British-type education and the mass media constructed our worlds as failed replicas of the modern West. This colonial effect of trying to learn from and imitate the global centre has been a preoccupation of post-colonial elites seeking to articulate a destiny that is a mixed set of western and Asian interests. Now a resident in the United States, my annual visits to South-east Asia intensify my awareness that an alternative vision of the future is being articulated, an increasingly autonomous definition of modernity that is differentiated from that in the West.

Aihwa Ong (1996: 60)

As Arjun Appadurai (1996) points out, much social science theorising of modernity has been concerned with explicating the modern as a distinct historical rupture. Diverse opinions as to precisely when this 'moment' occurred nonetheless share a common belief in the possibility of dividing the world not only temporally, but spatially too. A generalised opposition between 'the West' and 'the rest' sums up modernity's placing (Hall, 1992). The modern has, in many uses, been synonymous with the West. In this way, 'progress' in Malaysia (or elsewhere in the 'non-West') has been understood as merely mimetic, an act of 'replication', 'imitation' or 'catch up' (Ong, 1996). Whether couched in (once) optimistic vocabularies of 'development' or lamentations of cultural loss, a geography of modernity, in this light, implies little more than tracking the diffusion or expansion of an always-already modern West.

Three key theorists of modernity have, in different ways, (re)produced this Western-centred cartography of the concept. For Anthony Giddens, modernity refers to 'modes of social life which emerged in Europe from the nineteenth century onwards' (Giddens, 1990: 1). Modernity is a Western invention in that its 'two great transformative agencies' – the 'nation-state' and 'systematic capitalist production' – 'both have their roots in specific characteristics of European history and have few parallels in prior periods or in other cultural settings' (Giddens, 1990: 174–5). Certainly, Giddens has shied from positing an abrupt temporal shift to 'the modern'. What he terms the 'discontinuities of modernity' refer to the 'pace' and 'scope' of transformation in quantitative terms (Giddens, 1984: 4). Nonetheless, modernisation, in Giddens' scheme, remains a process of exporting Western institutions and formations to cultural settings of the non-West.

Michel Foucault also respected this spatialisation of modernity. In Foucault's work, modernity refers to a particular focus of power on the welfare of the population, the large-scale management of life and death. The development of 'bio-power' thus defined the 'threshold of modernity' for European societies (Foucault, 1990: 143). Yet, as Timothy Mitchell has pointed out, it was only by 'relegating the non-West to the margins and footnotes of his account' that Foucault was able to construct modernity as the story of Europe: 'To stage this homogenous time-space, there can be no interruptions from the non-West. The non-West must play the role of the outside, the otherness that creates the boundary of the space of modernity' (T. Mitchell, 2000: 15–16). Foucault, in other words, situated 'modern' governmental techniques squarely within the cultural history and geography of Europe.

Finally, for Bruno Latour, the modern has more literally been an historical 'invention' of 'the West'. What has been interpreted as 'the modern' is, in his view, an extension of scientific and institutional networks defining themselves as rational and true. Latour takes as his principal concern precisely how the 'Great Divide', which supposedly heralded and supported the modern, can no longer be sustained (Latour, 1993: 97). This unsustainability relates to the breakdown of two distinct but interrelated divisions: one which is internal, between culture and Nature; and another which is external, between 'moderns' (Westeners) and others (Latour, 1993: 133). However, David Matless (1996: 386) makes the critical point that 'the people of the nonwest seem to derive power in Latour's equation more from their exploding numbers than from any conscious thought and action'. Even in Latour's own critique of the modern, in other words, the effects and agency of the non-human do not appear to be matched by that of non-Westerners.

While these influential works – necessarily grossly abbreviated here – focus almost entirely on modernity of or as the West, they also identify key agents and processes for the study of other(s') spaces and places. Constellations of capitalist processes, the nation-state and bio-power extended the conceptual utility of 'modernity' beyond the imagined borders of (what is conventionally understood as) the West long before the current round of cultural-economic globalisation. Resultant transformation has involved – and continues to involve – a break from all kinds of 'problematic pasts' or, rather, from presents rendered irrational or traditional and so in need of 'development' (cf. Appadurai, 1996).[1] However, there is a danger here, in turn, of reducing change in the non-West to the effects of a (singular) modernity originating and authentically centred elsewhere. As such, in this chapter, I seek to elaborate not only how it is possible to speak of modernity in terms of the specificities of transformation in spaces beyond 'the West', but also: (1) how these transformations are animated by non-Western agency rather than reducible to effects of 'external' forces and processes; and (2) how such a conceptualisation might proceed without re-inscribing an imagined geographical opposition of the West and its 'others'. This clearly entails questioning the very binary spatial logics of internality/ externality through which modernity has largely been represented and understood.

To speak of modernity in this way is to consider its interrelated *contextuality* and *networked spatiality*. Work in historical geography is instructive here. As Catherine Nash and Brian Graham have pointed out, 'Historical geography has a long tradition of locating local studies within broader processes operating at wider spatial scales, of paying attention to both the specificity of the local and the wider economic, cultural and political processes and institutional structures' (Nash and Graham, 2000: 1). In part, this means attending to how geographies of modernity are shaped in and through context. However, as recent research has high-lighted, an imperative of contextuality is not reducible to ways in which putatively universal processes are differentially localised. Critical attention must be given to both the agency of local actors (Yeoh, 1996, 2000) and to what is meant by 'local'. In his work on eighteenth-century London's 'spaces of modernity', Miles Ogborn foregrounds the constitutive networks of connections between places rather than positing absolute locality. There is reciprocity here between local studies and broader processes: 'processes like individualisation, commodification and bureaucratisation need to be examined in the particular places that they make and that make them' (Ogborn, 1998: 19). Modernity's differentiated geographies, therefore, are not 'place-specific' but 'made in the relationships between places and across spaces'.[2] In this way, it is possible to reconceptualise transformations beyond the (post-)colonial metropole as less a matter of the transfer or imposition of a singular (and implicitly 'Western') modernity, than 'moments in the making of modernities' (*ibid.* 19).

The quotation from Aihwa Ong at the beginning of this chapter signals how we might take such reconceptualisation to the interconnected spaces of post-colonial Malaysia. Ong cast Southeast Asia in the 1990s as a site of 'increasingly autonomous' truth claims about the meaning of modernity itself. The West, in other words, was no longer imagined as the teleological end-point of being 'modern'. In the next section of this chapter, I seek to build upon existing work on modernity beyond 'the West'. On the one hand, I consider how elite geopolitical visions not only implied a discursive elaboration of 'Asian' development, but also served to legitimise particular socio-economic paths and cultural practices. On the other hand, I caution against conceptions of 'non-Western' modernity constructed through an imaginative geographical confinement of 'the West' which serve to invert existing cartographies of modernity. Drawing upon work on governmentality, the subsequent section then decentres political elite discourse (and 'the state') in emergent aims of and means to development. The 'developmental state' – so often a convenient short-hand for Asian or 'non-Western' socio-economic modernisation – is considered in relation to broader *rationalities of government* that make known (in)appropriate configurations of state–population–capital. Resultant national bio-political regimes are shown to be constructed relationally.

The remaining two sections of the chapter then consider the constitutive geographies of emergent modern regimes. The third section on 'new territories of government' is concerned with the spatiality of emergent regimes of government. In particular, while post-colonial states such as Malaysia have sought to extend development across their political territories as a means of nation-building, transnationalism and globalisation are shown to be bound up with a rescaling of state power and the fragmentation of the nation-state as a coherent territory of government. What Ong (1999) elaborates as the 'postdevelopmental state' renders visible new spaces and spatialities of government. Finally, the concept of *landscape* in cultural geography is introduced as a means of analysing spaces of government associated with emergent modernities in MSC–Malaysia. Landscape facilitates critical engagement with both the government of material micro-spaces and the multi-scalar discursive 'work' of modern technologies of visualisation.

Modernity and the 'non-West'

A range of studies increasingly take seriously the modernity of the non-West. Historical research has unsettled the genealogy of the modern by uncovering or recovering its emergence outside the imagined boundaries of the West. As Ann Laura Stoler (1995: 15) has noted, for example, students of empire have explored colonial sites as 'laboratories of modernity'. Stoler's own work extends the 'genealogical maps' of the development of European middle-class sexuality and selfhood beyond a European-bound geographical frame. Her critical re-reading of Foucault's *The History of Sexuality* traces how 'certain colonial prefigurings contest and force a reconceptualising of Foucault's sexual history of the Occident and more generally, a rethinking of the historiographic conventions that have bracketed histories of "the West"' (Stoler, 1995: 5). In so doing, Stoler historically problematises often taken-for-granted cartographies of modernity.

There is a danger here, however, that heterogeneous modern social arrangements emanating from the colonial 'margins' are simply folded into the historical emergence of a singular (Western) modernity. Thus, Timothy Mitchell (2000: 8) raises concerns about work in which 'developments outside "the West" are reorganised as part of its own history'. He focuses, in particular, on Sidney Mintz's (1985) *Sweetness and Power*, which considers how modern methods of industrial organisation were developed first in seventeenth-century Caribbean sugar plantations rather than for the manufacture of textiles in Manchester. Where Mitchell takes issue with Mintz is in his conceptualisation of Caribbean agro-industrial enterprises as merely 'nourishing' a stage of modern development 'at home': 'Historical time, in such an account, is singular, moving from one stage of development to another. There is no possibility of more than one history, of a non-singular capitalism.' (T. Mitchell, 2000: 9). This yields a geography of capitalism which does not allow for other spatialities and temporalities of modernity.[3]

Recent mappings of modernity in geography have also suggested possibilities for the concept's multiplicity. Peter Taylor's (1999) work expands modernity and its 'geohistory' so as to portray it as a 'global' phenomenon. On the one hand, Taylor tends to obscure disparate histories (and geographies) by foregrounding a consecutive series of hegemonic 'Prime modernities' (p. 28) in a (singular) 'capitalist world economy' (p. 24). On the other hand, however, he emphasises unequal relations of connectivity rather than mere homogeneity. Other geographical scholarship has given more explicit attention to 'alternate originations and evolutions' of capitalism (K. Mitchell, 1997: 104). The argument that modernity neither begins nor ends with the West has perhaps been most forcefully made by Allan Pred and Michael Watts. For these geographers, the self-mutation of capitalism 'has not only resulted in uneven development and a multiplicity of capitalisms, but precipitated a multiplicity of experienced modernities' (Pred and Watts, 1992: xii; see also Pred, 1995). Yet such critical reworking may, in turn, be problematised in two ways. First, as Timothy Mitchell has suggested, the language of alternative or multiple modernities is an 'easy pluralism' merely positing variants of underlying fundamental processes (T. Mitchell, 2000: xii). Second, and relatedly, such pluralism reduces variation to 'local' cultural resistance to, or modification of, supposedly more fundamental global (political economic) processes (Ong, 1999).

Both of these strands of critique demand that we think critically about an analytical distinction whereby 'the global is macro–political economic and the local is situated, culturally creative, and resistant' (Ong, 1999: 4). It is, of course, established practice for social scientists to seek to grasp *local* conditions and perspectives thereby recognising 'local' actors as producers of local cultural knowledge. However, a tendency to 'indigenise knowledge' serves to define and confine such knowledges as 'absolutely local, without scope or wider application' (Moore, 1996: 6).[4] Modernity's multiplicity is not to be conceptualised in terms of an 'easy pluralism' of contextualisation through which: (1) an implicitly Western 'global' is localised in 'other' places; or (2) a transcendent 'political economic' is modified in relation to diverse 'cultural' knowledges. Rather, to recognise the modernity of other(s') places is to acknowledge the non-West as a source of self-theorisation and truth claims – the non-West as producer, as well as mediator, of knowledge which is extra-local, even global in scope.[5]

Such 'self-theorisation' has been evident in Southeast Asia in the form of 'increasingly autonomous' alternative visions of modernity, particularly during the 1990s. As sketched in the previous chapter, confidence borne of prolonged economic success and relative social and political stability emboldened political leaders in states such as Malaysia and Singapore to 'break discursively with Western ideals of the modern' (Ong, 1999: 59). 'Asian values' and 'Asian Renaissance' became geopolitical banners under which elites constructed truth claims about their own countries and/or regions. Thus, on the one hand, as Ong suggests: 'we need to attend to how places in the non-West differently plan and envision the particular combinations of culture, capital and the nation-state, rather than assume that they are immature versions of some master Western prototype' (*ibid*.: 31).

On the other hand, we must be wary of positing unified 'alternative modernities' that reify a West–non-West dichotomy. Certainly, self-confident political envisionings of 'alternative' futures by Asian leaders in the 1990s challenged assumptions of Euro–American socio-economic leadership. Yet 'Western' contributions to the 'Asian' present were also deliberately downplayed in 'contexts of shifting geopolitical alignments' (*ibid*.: 53). The discursive strategies of states in Southeast Asia served to invert Euro-centred cartographies of modernity: a culturally distinct Asian modernity was rendered visible through the spatial confinement of the 'Western Other' (Yao, 2001b: 64). It is important not to lose sight of the fact that these envisionings of appropriate, modern futures – and means to their realisation – emerged from (post-)colonial geo-histories of interconnection, not only in the 'economic' domain, but also in terms of the circulation of politico-cultural ideals and practices.

This is not to suggest that modernity's multiplicity extends no further than discursive reworking emanating from political elites or 'the state'. Without doubt, often unsophisticated visions of 'development' in Southeast Asia have been made possible by 'raiding the rich storehouse of Asian myths and religions' (Ong, 1999: 7). Yet even the recognition that much such discourse – in Malaysia at least – is 'remarkably thin, highly repetitive, sometimes inconsistent and even self-contradictory' is not to deny its significance in legitimising capitalist strategies and particular forms of conservative social and economic modernisation (see Kahn, 1997: 19). In addition, as C. J. W.-L. Wee has demonstrated, normative culturalist discourses have socio-economic (as well as political) effects. 'Statist Asian values discourse' has been actively bound up with the management of customs and culture for global capitalism (Wee, 2002: 9). These are strong motivations for a post-structuralist geographical critique of spatially- and culturally-essentialist state discourse. Another motivation, of course, is the by now 'conventional theoretical wisdom that the discursive realm is never purely "representational", but has emerged from and consolidated into real power by legislative framing and legal enforcement' (Yao, 2001a: 4).

The extent of such political power in much of East and Southeast Asia has often been expressed in the generalised regional label of 'developmental state'. On the one hand, the 'fundamental presence' (Yao, 2001a: 6) of state power in Southeast Asia, as distinct from that in many other regions, is undeniable.[6] On the other hand, however, it is important to acknowledge that state discourse and practice – in Southeast Asia as elsewhere – is itself an effect as well as a cause of spatially- and temporally-specific constellations of political and cultural forces. It is to a political decentring of the so-called 'developmental state' which I now turn.

The developmental state governmentally reconsidered

A number of states in East and Southeast Asia have been termed 'developmental' on account of their supposedly 'strong role' in economic and social modernisation (Johnson, 1982; Wade, 1990).[7] State-driven capitalist development in Asia is evidenced from attempts to promote local enterprises through state subsidies and protection as well as from the formation of alliances with domestic and transnational capital that have motivated both state disciplining and repression of labour and the fostering of technical and professional middle classes. While this has been differentiated from modern Anglo-American-led neo-liberal regimes, the developmental state, I argue, is not an adequate concept for understanding social and spatial transformation in the changing landscapes of Malaysia's Multimedia Super Corridor.

The conceptual inadequacy of the 'developmental state' may be understood in three ways. First, and most simply, there is great diversity among nation-states classified as 'developmental' (Trezzini, 2001). A generalised regional concept has doubtful utility for a book seeking, as this one does, to bring in contextual variability and alternative spatialities of modernity. Second, and as already alluded to above, state power is itself an effect of broader forces. Attention must be given to the intertwined cultural and political contestation out of which conceptions of appropriate 'development' emerge – including the role of the state. Third, and relatedly, transformative agency, even in so-called developmental contexts, extends beyond state power. In this research, I consider socio-cultural and economic (self-)management associated with various authorities including, but not only, state actors. Here I draw upon the burgeoning governmentality literature[8] in seeking to decentre the state from 'political' analysis.

Michel Foucault, in his seminal essay, *Governmentality*, highlighted analyses of political power that were not transfixed by the image of the state:

> The state, no more probably today than at any other time in history, does not have this unity, this individuality, this rigorous functionality nor, to speak frankly this importance: maybe, after all, the state is no more than a composite reality and a mythicised abstraction, whose importance is a lot more limited than many of us think. Maybe what is really important for our modernity – that is, for our present – is not so much the *étatisation* of society, as the governmentality of the state.
>
> (Foucault, 1991: 103)

Government, in other words, extends beyond what we conventionally understand by the political. The 'governmentality of the state' refers to a broader conception of political power as action upon the details of the conduct of individuals and population (the 'conduct of conduct'). In relation to this objective, as Nikolas Rose (1999: 5) puts it: 'The state now appears simply as one element – whose functionality is historically specific and contextually variable – in multiple circuits of power, connecting a diversity of authorities and forces, within a whole variety of complex assemblages.' While not denying the 'massive capability of the state in Southeast Asia and its permeation of the social, economic and cultural life in the region' (Yao, 2001a: 6), governmentality helps us to see political power beyond 'the state' (and especially the 'developmental state').

Two interrelated conceptual imperatives arise from a governmental reframing of the state and its relation to 'political power'. First, attention should be accorded less to the state in and of itself than to the conceptualisations and mentalities, contests and debates which make known appropriate government. Government here includes, but is not reducible to, the role of the state. State (in)action, therefore, is merely part of broader questions about who or what should govern and be governed, by what means and to what ends. Second, this in turn suggests that we extend research from the activities and intentions of the state to also include consideration of the governmental work of non-state actors, forces and institutions. In an influential contribution which drew upon *Governmentality*, Peter Miller and Nikolas Rose (1990) highlighted the diversity of forces and groups – many of which are only loosely associated with the executives and bureaucracies – which seek to regulate the lives of individuals and the conditions within national territories. In this way, government is no longer understood as synonymous with 'political authorities' in the conventional sense, but rather refers to 'the diversity of power and knowledges entailed in rendering fields practicable and amenable to intervention' (Miller and Rose, 1990: 3). What Foucault (1990) termed 'bio-power', then, does not refer merely to state power. Rather, it points to the diversity of authorities and strategies oriented to the enhancement of the lives and welfare of populations in different national territories.

The governmentality literature has overwhelmingly considered 'liberal' regimes for the conduct of conduct that have taken shape in 'the West'[9] raising the important question of its applicability to other spaces of modernity. The classic dilemma for liberal modes of government is how 'free' individuals, groups and organisations can be governed 'such that they enact their freedom appropriately' (Rose, 1996b: 134). Governmentality, however, is not merely diagnostic of societies that might be identified as liberal and those which are not; nor, in fact, does it hold up 'the liberal' as a 'minimal state' end of a spectrum of state power (Dean, 1999: 101). Rather, Mitchell Dean, for example, has cast liberalism as an 'ethos of review' whose problem 'will be not a negation of bio-political regulation but a way of managing it' (*ibid*. 101). Even in supposedly 'developmental' contexts, conceptions of the appropriate role of the state are reviewed, negotiated and reworked. And even in authoritarian states, non-coercive forms of power work through domains that are more or less free from direct regulation.[10] What governmentality highlights, in other words, is how political rationalisation shapes the conduct of the state and its relation to other authorities and domains. In a similar way, in her work on East and Southeast Asia, Aihwa Ong has considered different 'modes of biopolitical regimes' – particular configurations of state, population and increasingly global capital – within national territories (Ong, 1999).

The national scale territorialisation of bio-political regimes, however, also demands critical elaboration. In the first place, political rationalisations are clearly (re)shaped by extra-national discourses and conceptualisations. On the one hand, as Nikolas Rose has suggested, Foucault's (1991) work on governmentality, 'implied that one could identify specific political rationalisations emerging in precise sites and at specific historical moments' (Rose, 1999: 24). It is important to take seriously 'contextual variability' in regimes of bio-political management. On the other hand, such variability should not be taken to imply bounded topologies of 'the political'. Apart from the networked spatiality of modern processes referred to already in this chapter, discourses of governmentality – such as supranational organisations' association of liberalisation in various domains with 'good governance' (Ó Tuathail *et al.*, 1998) – are potentially global in scope. Recent work in geography on scale (Kelly, 1999) suggests a *relational* resolution of such issues. National scale political rationalities are understood to be (re)constructed in relation to discourses and conceptualisations at other scales.

A second point concerns the coherence of the nation-state as a bio-political territory. Contemporary processes of globalisation and transnationalism mean that the spatialisation of bio-power at the national scale is itself in question. At one level, this relates to the way in which the nation-state has been 'de-naturalised' in the social sciences (see Taylor, 1996). For Euro-centred social science, it was in part precisely the globalisation of the social world under study which compelled an escape from the 'territorial trap' of state-centric thinking (Brenner, 1999). Thus, according to Nikolas Rose, conventional ways of analysing politics and power now appear increasingly obsolescent as they are associated with a period 'when the boundaries of the nation state seemed to set the natural frame for political systems' (Rose, 1999: 1). However, the apparent need for such a reframing is also a result of the 'real' transformation of state power and its relation to broader bio-political regimes. One key geographical aspect of the qualitative reorganisation of government – including the role of state power – is its increased spatial differentiation. It is to the respatialisation of political strategies and the emergence of new territories of government which I now turn.

New territories of government

Analyses of political power in a time of accelerated globalisation and transnationalism have largely centred upon the putative quantitative erosion of state capacities (see, for example, Ohmae, 1992, 1995). While long-distance connections forged by trade and migration are far from new, contemporary transnational networks associated with advances in transport and communications technologies have greatly increased possibilities for cross-border human mobility and capital flows. Without doubt, the nation-state, traditionally regarded as the key unit of social and economic organisation as well as the central building block of the geopolitical system, has been impacted profoundly by the rise of transnational processes and imaginings. If conceptions of a shift from 'nations to networks' suggest a qualitative diminution of the power of the former at the expense of the latter, however, a range of recent work has pointed rather to the qualitative reorganisation of the strategic emphases and capabilities of the nation-state (see Dicken *et al.*, 1997; Yeung, 1998). This qualitative reconfiguration, I suggest, has an important geographical dimension. What James C. Scott has referred to as states' way of 'seeing' their population and territory (Scott, 1998) is increasingly spatially as well as socially differentiated. Ong associates this with what she terms the 'postdevelopmental state' (Ong, 1999: 21). While this term perhaps implicitly gives overdue unity to the state form which it ostensibly follows – the developmental state – Ong's work is very fertile ground for new geographies of government.

A useful starting point here is to consider work on the 'rescaling' of state power. Much has been made of the rise to prominence of cities and urban regions in globalising times (see, for example, Scott *et al.*, 2001). On the one hand, some have argued that this rise signals a corresponding decline in the economic, political and social significance of nation-states (Ohmae, 2001). Cities, not nation-states, are commonly understood as the key actors of the information society and economy (Castells, 1989, 1996). Compared to national governments, cities are considered to be altogether 'more flexible in adapting to the changing conditions of markets, technology and culture' in a so-called 'Information Age' (Castells and Hall, 1994: 7). On the other hand, however, rapid urban-regional development need not be antithetical to state power. In many cases such development is an expression of re-scaled state power. Thus, recent urban research has suggested that, 'currently unfolding re-scalings of state institutions signal not the decline or erosion of the state, but rather a specifically geographical accumulation strategy to promote and regulate industrial restructuring within major urban regions' (Brenner, 1998: 476).

These political strategies have been rationalised by broader discourses of urban and regional development. In the late 1990s, key supranational policy documents such as the World Bank's (1999) *Global Urban and Local Government Strategy* promoted the devotion of greater national attention and resources to key city-regions. More pervasively, a global diffusion of consultants, real estate interests and Transnational Corporations (TNCs) imagined the production of 'intelligent' urban spaces as an essential ingredient for global economic success (Graham and Marvin, 1999). A 'global' or 'world' city was imagined as a national means of 'plugging in' to global political, economic and social networks (Robins, 1999). Despite the Anglo-American-centredness of much of the 'international' literature on urban and regional research, such trends have also been evident in the Asia-Pacific region (Olds, 1995, 2001; Lo and Marcotullio, 2000).[11] According to Mike Douglass (1998: 111), the active intervention of Pacific Asian governments in urban restructuring has been a response to the belief that, 'the status of their national economies will be increasingly determined by the positioning of their principal urban regions'. Clearly this entails and effects a reworking of conventional conceptions of 'national development'. In Southeast Asia, as elsewhere, nation-states are abandoning aspirations of 'homogenising spatial practices on a national scale' (Brenner, 1998: 476; see also Bunnell, 2002b). Partly in relation to discourses and practices of globalisation, states in the region are re-envisioning their territories in new or, more specifically, newly-differentiated ways.

I suggest that this re-scaling forms part of a broader break-down of (bounded) national territories of government, leading us back to the work of Aihwa Ong. What is really important, she suggests, is the way(s) in which the hyphen in 'nation-state' has become reconfigured by capital mobility and migration. Again, this is not to suggest a wholesale diminution of state capacities, but is perhaps more accurately understood in terms of the emergence of new spatialities of bio-political power. In Ong's (1999: 15) words, the state 'continues to define, discipline, control and regulate all kinds of populations, whether in movement or in residence'. In some cases, this has entailed extending the 'national project' beyond the bounds of the national territory to include citizens overseas. Michael Peter Smith has noted how, 'many nation-states that have experienced substantial out-migration are entering into a process of actively promoting transnational reincorporation of migrants into their state-centred projects' (Smith, 2001: 172). This may be in order to attract back those nationals who have acquired valuable skills, to secure their return investment, or else to capitalise on them as conduits to extra-national economies. What is of more general interest here, however, is that certain segments of the 'national' population are increasingly valued (while others are not). People (and places) are (re)incorporated into state-centred projects in increasingly uneven ways. There is, in other words, growing socio-spatial differentiation of the population in relation to the power of the state and other authorities.

Aihwa Ong elaborates governmental reconfigurations in relation to the emergence of her conception of the 'postdevelopmental state' in Asia.[12] In terms of state strategies, this refers to an increasingly 'unequal biopolitical investment in different categories of the population' (Ong, 1999: 217). Individuals and groups may be subjected to different regimes of value on account of their gender, ethnicity and/or human capital and skills. Attempts to produce society in ways that are attractive to global capital also relate to modes of (non-)regulation along class lines. The middle classes are increasingly subjected to non-repressive modes of cultural government to realise themselves in ways that will foster faster development. Meanwhile, 'weaker and less-desirable groups are given over to the regulation of supranational entities' (*ibid.* 217). There thus emerges a system of 'graduated sovereignty' which sees 'the population' divided into 'different mixes of disciplinary, caring and punitive technologies' (*ibid.* 217). New modes of bio-political regimes associated with these post-developmental shifts between governmentality and sovereignty suggest a further fragmentation of the nation-state as a territory of government.

Thinking geographically about this fragmentation allows us to reframe the re-scaling of state power in terms of a broader emergence of new territories or spaces of government. There is a clear geographical dimension to Ong's conception of graduated sovereignty: 'Different production sites often become institutional domains that vary in their mix of legal protections, controls and disciplinary regimes' (*ibid.* 215). Ong, in fact, uses Malaysia to illustrate the existence of 'zones' of graduated sovereignty within or (as in the case of regional growth triangles) across the geo-political boundaries of the nation-state.[13] The Multimedia Super Corridor (MSC) is thus cast as a 'zone of superior privileges' for the realisation of 'a new kind of Malay subject who will be fully at home in a multimedia world' (*ibid.* 219). The reconfiguration of state power associated with the valorisation of high-tech urban spaces – and suitably high-tech citizens – is thus understood in terms of emergent post-developmental zones of government.

However, a critical geographical analysis of spaces of government in MSC/Malaysia implies more than a simple identification or mapping of variegated zones. Two issues, in particular, are worthy of note here. The first concerns what might be understood as the productive or constitutive role of space (see Harvey, 1989; Lefebvre, 1991). That is, regimes of government do not merely occur in and/or vary across space. Rather, space is actively bound up with the 'conduct of conduct' – in the case of this book, the realisation of would-be 'intelligent' subjects. The second issue concerns the political tensions of differentiated spaces and scales of government. In Malaysia, as elsewhere, the federal government privileging of key zones sits uneasily alongside an electoral mandate at the national scale. More specifically, socio-spatially variegated post-developmental state practices appear to subvert a post-independence nation-building project ostensibly oriented to spreading 'development' more uniformly across the national territory. These tensions will be covered in more detail in the next chapter. In the final section of this chapter, I consider how the concept of landscape in cultural geography may be used to analyse the governmental work of MSC space and its scalar tensions.

The government of landscape

I consider the concept of landscape here as a way of thinking critically about spaces of government associated with emergent modernities of MSC/Malaysia. Landscape, as it has been understood in the so-called 'new cultural geography', refers to both material space and ways of seeing/ representing space in various media.[14] Over the past two decades, cultural geographers have embraced methods developed in art history (see, for example, Daniels and Cosgrove, 1989) and literary theory (Duncan and Duncan, 1988) in seeking to analyse landscapes. Landscapes, 'real' and/or 'representational' can thus be read as texts (Duncan, 1990; Donald, 1992; Goh and Yeoh, 2003). In part, this book follows such approaches, interpreting the high-tech landscapes of MSC by re-immersing them in the geo-historical contexts of their production and by 'peeling off' or uncovering layers and levels of meaning (Domosh, 1989). However, I also seek to combine these interpretative approaches with work on non-representational theory to consider the constitutive *effects* of landscape. Here I thus elaborate what I term the 'government of landscape'.

Non-representational theory[15] has exposed some limit(ation)s of interpretative landscape approaches. Non-representational retheorisation is based on the argument that representations are not simply to be 'read' by more or less skilful social scientists revealing how the world 'really is' (Thrift, 1997). Rather, representations are to be approached *practically* and *relationally*, attending to the role they perform, their 'effects' (Thrift *et al.*,

1995). This applies equally to material landscapes which are thus not merely signs or clues to understanding pre-existing politico-cultural formations or geo-historical structures, but bound up in the unfolding of ongoing transformation. More than this, in a recent article, Nigel Thrift (2000a: 689) has argued that, 'standard oppositions of the kind – lived and represented, experiment and conceptualised, abstract and concrete – are unhelpful, even actively misleading'. In disciplinary terms, this would appear to suggest a challenge to what he casts as 'an increasingly predictable and anaemic hegemony of cultural geography' (Thrift, 2001: 556) uncritically content to seek to turn the 'world into the word' (Thrift, 2000b: 1).

I suggest that it is already possible to incorporate non-representational critique as part of cultural geography's own ongoing (re)development. Cultural geographers themselves increasingly work through non-representational strategies rather than focusing solely on more established concerns for 'text' and 'representation'. David Matless (2000: 336), for example, has carried out a geographical examination of a range of objects 'less through an assumption that hidden relations are concealed in a finished form which thereby requires dismantling, than by considering that finished form as one significantly congealed state with a wider field of relations of which it is an effect'. Material objects, in other words, are bound up with 'geographical practices which go to make up senses of the self' and 'the production of geographical subjectivities' (Matless, 2000: 335–6). It is in this way that landscape has been understood as a 'quasi-object' (Matless, 1998: 12) with 'real' effects. It may in fact be argued that cultural geographers have long acknowledged how landscapes are *active* in shaping individual and collective socio-cultural practices. Social and political authorities have been shown to (re)define norms, ideals and objectives through landscapes in turn inducing new 'ways of seeing'. Textual, pictorial and other representations, non-material landscapes, thus 'naturalise' social aims and ideals (Daniels, 1988: 43). Landscape, both discursive and material, has been understood not only as a noun, but also as a verb (Mitchell, 1994).

Recent work on landscape in cultural geography has made explicit how landscape may be understood as both a 'work' and as something that 'does work' (D. Mitchell, 2000: 94). The latter might conventionally be understood in terms of the use of architecture and building design as technologies of government, fostering modern practices of working (Gottman, 1966) and living (Dennis, 1994; and see also Markus, 1993). Yet what I term the 'government of landscape' extends beyond the 'work' of physical design. Landscape studies in cultural geography have considered the role of various authorities in shaping conceptions of appropriate 'land use' and environmental conduct (see, for example, Matless, 1994; 1998). The active role of space in the constitution of such conduct – the governmental work of landscape – is thus, in part, *normative*: the way in which ideas about what is right and appropriate are 'transmitted through space and place' (Cresswell, 1996: 8). Individuals sustain and remake themselves through their environment, which they in turn reconfigure in geographical 'practices of the self' (Matless, 1994; and see Foucault, 1988). An important geographical point here is that the contested emergence of authoritative systems of evaluation has a more complex spatiality than is captured by 'local' or site-specific 'cultural politics'. Richard Schein has usefully conceptualised landscape as 'an articulated moment in knowledge networks stretching across spaces' – landscape as 'node' (Schein, 1997: 662–3). Landscapes are thus extra-locally constituted sites and sights that in turn reconstitute geographical selves. Nodal discursive-material assemblages are spaces of government for new landscaped subjectivities.[16]

The governmental work of landscape extends beyond micro-spaces or specific nodes. The 'lure of the local' in cultural geography (Mitchell, 2001) has not precluded analyses of the work of landscape at the level of whole (national) populations. Landscapes in various media have been shown to articulate national identity (Daniels, 1993) which is, in turn, reworked and reconstructed through practices of individual and collective consumption (Johnson, 1995). As I seek to demonstrate in subsequent chapters, MSC landscapes are powerful 'technologies of nationhood' (see Harvey, 1996: 56) making known authoritative aims of and means to national development. Idealised national landscapes demonstrate appropriate or exemplary, individual and collective conduct which is folded into the (self-)regulating judgements and calculations of citizens. Of course, there are vast and apparently increasing socio-spatial disparities in the likelihood (and desirability) of realising these 'national' subject ideals. As suggested in the previous section, MSC has been cast as a 'superprivileged zone' of government differentiated from regimes of bio-political power in much of the rest of the Malaysian national territory (Ong, 1999: 220). The 'government' of MSC landscapes thus encapsulates tensions implied by the project as at once a symbolic national node and a distinct bio-political zone which diagnoses and contributes to the post-developmental fragmentation of the national territory of government.

It is important to make two critical qualifications to the generalised conception of the 'government of landscape' outlined so far. The first concerns the *spatiality* of different governmental practices. While proponents of governmentality have argued for attention to be paid to variable forms and practices of power, geographers have highlighted, in particular, the role of geographical *scale* in such variation (Hannah, 1998; Barnett, 1999). Clive Barnett (1999), for example, has argued that the regulation of micro-spaces must be distinguished from the administration of whole populations. The need for such a distinction addresses, in part, a rather different conflation – between disciplinary and governmental power – but clearly landscaped practices of the self in newly-fashioned 'intelligent' urban spaces are to be distinguished from 'government at a distance' associated with landscape imagery disseminated through media and communications technologies.[17] Barnett, in fact, goes on to make a more complex argument about the reconfiguration of the spaces of government associated with radio and television, positing a need for a 'networked conception of socio-spatial power' (Barnett, 1999: 389). At the very least, however, it is clear that the successful translation of authoritative ideals into individual and collective practices at the national scale is likely to be highly socially and spatially uneven.

This leads directly onto the second qualification which concerns the *effectiveness* of governmental strategies. While the efficacy of governmental strategies is certainly variable – not least in relation to issues of scale and topology alluded to above – government, more generally, is a 'congenitally failing operation' (Miller and Rose, 1990: 18). This relates, in part, to Foucault's reconceptualisation of 'the subject' in his work on governmentality. Critical arguments against Foucault's radical decentring of the subject are well rehearsed (see Hall, 1997), but it was precisely through the concept of governmentality that he sought to overcome these in his later work (Foucault, 1982; and see McNay, 1994). The theme of governmental power is to be distinguished from earlier concerns with the formation of domains of knowledge and with punitive rationalities in its concern with 'rehabilitating agency' (Barnett, 1999: 383). As Mitchell Dean puts it, 'Government concerns the shaping of human conduct and acts on the governed as a locus of action and freedom. It therefore entails the possibility that the governed are to some extent capable of acting and thinking otherwise' (Dean, 1999: 15). The on-going failure and problematisation of practices of government gives rise to a reworking of rationality which is itself a source of future political inventiveness (Dean, 1994). The governmentality literature thus not only highlights ongoing contestation, but also how this (re)constitutes rationalities through which (self-)government takes place.

'Contest' here is to be distinguished from social scientific conceptions counter-posing power and its opponents, radically or romantically marked 'resistance' (see Thrift, 1997). As an 'applied art' (Dean, 1994: 187), government is inevitably bound up with contest and competition among individuals and groups. However, following Foucault, it is misleading to see authorities as constitutive of 'power' acting upon and/or resisted by everyday individuals and groups. Rather, we should seek to chart the contested (re)construction of appropriate conduct and means to achieving it. Similarly, trangression does not so much oppose or resist power, but is both 'diagnostic' of hegemonic socio-spatial practices (see Cresswell, 2000) and a spur to new governmental innovation (Bunnell, 2002c). In sum: 'our present has arisen as much from the logics of contestation as from any imperatives of control' (Rose, 1999: 277). This includes the landscape, which is shaped by and active in governmental contestation.

Finally, it is precisely through foregrounding the governmental 'work' of landscape that we can begin to build a critical cultural geography of MSC. Who belongs in and who is excluded from utopian MSC spaces of government? Which individuals and groups are (un)able to realise themselves in 'intelligent' ways through high-tech urban landscapes? These two interrelated questions point to emergent 'dividing practices' (see Rose, 1996b: 145) associated not only with the material and discursive construction of MSC but also ultimately with the broader conceptualisations and calculations which rationalise the project. At one level, MSC landscapes may be 'read' in terms of a fragmentation of the national territory of government. In addition, as I have sought to signal here, landscapes of MSC effect new social and spatial divisions. The material and discursive landscaping of Malaysia(ns), as I will demonstrate in the chapters that form Part II of this book, is characterised by new socio-spatial dividing practices at a variety of scales.

In this chapter, I have examined possibilities for understanding modernity beyond the imagined territory of 'the West'. Other(s') places have not only been profoundly implicated in the geo-historical 'development' of Western Europe and North America – and of 'the West' that has conventionally been seen as monopolising the leading edge of modernity – but, perhaps more explicitly than ever in Southeast Asia in the last decade of the twentieth century, they envisioned and effected 'progress' in putatively autonomous ways. Modernity may, therefore, be understood in and through specific places and spaces rather than as an exported set of essentially Western processes and institutions. I have suggested that we conceptualise alternative experiences of modernity in terms of geo-historically specific political rationalisations – aims of and means to development, including the role of the state. However, I have also highlighted two critical points about the spatiality of modernity's multiplicity. First, rather than positing bounded 'alternative modernities', place and politics are understood to be constructed relationally. Contested conceptions of national development are forged through networked interconnection. Second, emergent rationalities associated with globalisation and transnationalism imply the fragmentation of the nation-state as a coherent or unified territory of government. I have proposed the concept of landscape in cultural geography as a means of analysing Malaysia's MSC as at once a site/sight for intelligent subjectification, a technology of nationhood and a space of government which diagnoses a broader national socio-spatial fragmentation.

Lastly, it is important to note that while in Euro-American centred social science processes of globalisation and transnationalism may have served to reveal the national 'territorial trap' in which much research had been ensnared, such 'theoretical developments' perhaps appear less novel or significant in former colonial contexts where nation-states have never been naturalised or taken-for-granted. Nonetheless, it is clear that globalisation and transnationalism are bound up with the ongoing fragmentation of 'the political' such that uneven experiences of modernity defy reduction to a single moment or spatial divide. It is to the geo-historical construction of Malaysia and its ongoing global(ising) reconfiguration that I turn in the next chapter.

3 Positioning Malaysia
Connections, divisions and development

'The dominant images we have of Malaysian society – whether they are conveyed to us in the sociological literature, the speeches of politicians, the brochures of the tourist industry, or the everyday discourse of ordinary Malaysians – are images of diversity' (Kahn, 1992: 158). So begins Joel S. Kahn in an essay on the (re)construction of 'Malayness' in contemporary Malaysia. The 'Malays' are the largest of the three main official communities which comprise 'plural' or 'ethnically-divided' Malaysian society, the other two being the 'Chinese' and 'Indians'. To the extent that Malays have lived traditionally in rural areas and many non-Malays in towns and cities, it is possible to identify a historical geography of mutually-reinforcing spatial and ethnic divides. Malaysia is also geo-physically divided: between the eleven states of Peninsula Malaysia, on the one hand; and two East Malaysian states (Sabah and Sarawak), on the other (see Figure 3.1). Division and difference are clearly deserving of attention in any attempt to contextualise socio-spatial transformation in contemporary Malaysia. Presumptions and constructions of diversity have framed contested political notions of what the nation could (or could not) be, and what 'development' should (or should not) be.

Colonialism figures prominently in accounts of both the making of Malaysia and its fragmentary characteristics. On the one hand, British colonialism brought a measure of economic and administrative unity to previously disparate and often disputatious territories, especially in the Peninsula. Colonial Malayan economic spaces were integrated into a recognisable urban system (Lim, 1978). On the other hand, however, the division of labour which emerged to service the colonial economies of Southeast Asia cemented ethnic or 'racial' differences (Furnivall, 1939).[1] Each group had their respective places in colonial economies which thus re-diagrammed existing social and spatial divisions. Contradictory colonial logics of integration and (re-)division are clearly indispensable to any introduction to (post-)colonial Malaysian development and they inform much of this chapter.[2] However, I raise two further points about the conceptualisation of colonialism in this geo-historical *positioning* of Malaysia.

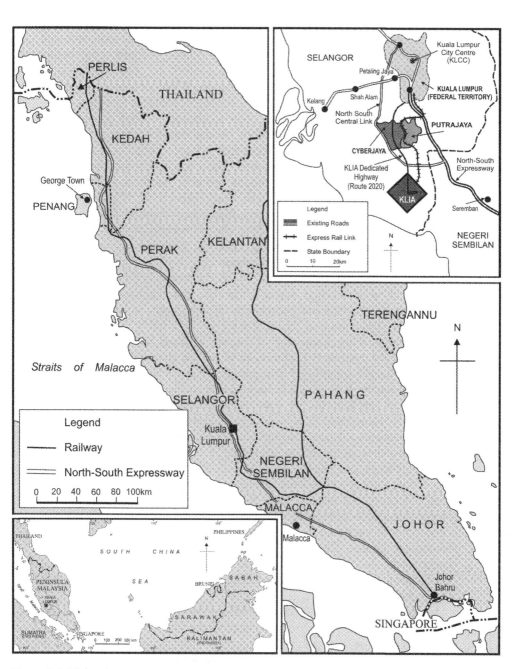

Figure 3.1 Malaysia.

The first concerns an imperative of decentring colonialism (see Dirlik, 1994). Formal colonisation by Britain in the nineteenth century followed more than three hundred years of European commerce, conquest and competition in Southeast Asia. More importantly, perhaps, long before Portuguese expansion in the late fifteenth century, the lands and seas of present-day Malaysia were bound up in geographically-extensive maritime trade (Reid, 1992; Wang, 2001). The port town of Malacca (see Figure 3.1), captured by the Portuguese in 1511, was a culturally-diverse entrepot with a population of as many as 100,000 (Andaya and Andaya, 2001: 46). Malacca town's nodal position meant that it was enrolled in cultural and religious as well as economic networks. Yet it was economic interests and opportunities, above all, that brought growing numbers of sojourners and migrants from China, India and the Hadramaut to the Malay Peninsula in particular, again long before any formalised colonial economy (see Wheatley, 1961). This is not to deny the transformative significance of British colonialism, but to recognise that this needs to be positioned in relation to other constitutive geo-histories.

This leads on to a second point which concerns the implications of a diversity of geo-historical forces, interests and agencies for colonial administrative practice and power. On the one hand, 'colonialism' itself has perhaps been granted undue coherence and unity. Albert Lau has cautioned against any suggestion of colonial ideology in the singular (Lau, 1991). Critical historical accounts now accept that while the pluralist system in Malaya functioned in economic terms, it was only experiences during the Pacific War which brought home to the British administration the political implications of communal division (Watson, 1996). The bureaucratic formalisation of racial groups and the economic spatial divide that concretised such social divisions have perhaps been attributed undue organisational capacity and intentionality in notions of 'devise, divide and rule' (see Cleary and Shaw, 1994). On the other hand, it is important to acknowledge other voices and practices in shaping Malay(si)a. Shamsul A. B. has examined the historical role played by popular resistance and protest during the British colonial period (Shamsul, 1986). Geographical work in Southeast Asia too has focused upon issues of everyday contest to colonial strategies. Brenda Yeoh's work on Singapore, for example, foregrounds the agency of indigenous and non-colonial immigrant populations in shaping and defining space in the colonial urban environment (Yeoh, 1996).

These critical reworkings of colonialism suggest important lessons for analyses of more recent socio-spatial transformation. In the first place – and as considered in the previous chapter – the complex spatiality of transformative processes and agencies unsettles binary logics of internality and externality. Just as social transformation and commercial development in the Malay Peninsula pre-dated the arrival of a supposedly modern(ising) Europe, Malaysian national development has been characterised by, and remains, a mixed set of constitutive extra-local relations. In this chapter, in addition to material colonial (infra)structures and legacies, we variously encounter supranational organisations' prescriptions for modernisation, Social Darwinian or neo-Lamarckian evolutionary biology, East Asian developmentalism and neo-liberal discourses of globalisation in relation to, and *as part of*, Malaysia's post-colonial landscapes. Recognition of the hybrid (re)constitution of Malaysia's modernities became all the more important from the 1990s as politically powerful conceptions of 'the way forward' (see Mahathir, 1993) were articulated in increasingly essentialist national and/or regional terms.

A second lesson that may be learned from a critical repositioning of colonialism concerns the way in which we conceptualise 'political power'. Critique of presumptions of the coherence and organisation of colonial power connect to a broader decentring of conventional political authorities detailed in the previous chapter. The state, its institutions and personalities are thus understood in relation to broader *political rationalities*. Analysis at this level is perhaps particularly important for the last two decades of the twentieth century in Malaysia when economic and social transformation – whether viewed as being for better or worse – has frequently been attributed to the actions, idea(l)s and vision of its Prime Minister. One popular evaluation of Dr Mahathir Mohamad suggested that, 'what he thinks and how he thinks has permeated the fabric of society, colouring it much like a dye, until the society resembles him' (Karim, 1996: 122). While Mahathir figures prominently in this chapter – and in the book as a whole – it is clearly unsatisfactory to understand the modernity of a particular time and place in terms of an omnipotent leader somehow transcending history, geography and politics. Writing in the early 1980s, Roger Kershaw suggested that it would be 'most meaningful to see the Malaysian Prime Minister as a man who responds to the inchoate visions of a new Malay intelligentsia as much as he moulds them' (Kershaw, 1982: 647). The genealogy of what Khoo Boo Tiek has termed 'Mahathirism' in fact extends much further than this, both historically and geographically (Khoo, 1995: 2). Transformation is not reducible to the will of the state or its key personalities to colour society and space in their own image.

In the remainder of this chapter, I detail four broadly chronological stages in the (re)positioning of Malaysia. The intention is not to impose some kind of singular narrative upon Malaysian history which sees Vision 2020 and the Multimedia Super Corridor (MSC) as a distinct and inevitable final or cumulative stage. It is precisely by thinking about socio-spatial transformation through ongoing reworking of difference that I hope to avert this danger. To acknowledge presumptions of difference is, in part, to acknowledge multiple historical visions of progress and, ultimately, the socio-spatial contest and fragmentation which remain part of emergent modernities of contemporary Malaysia.

Making Malaysia(ns)

The Federation of Malaya gained independence from Britain on 31 August 1957, but it was another eight years before the Malaysian nation-state came to occupy its current physical territory. In 1961, Prime Minister Tunku Abdul Rahman proposed the idea of a 'Grand Malaysian Alliance' which would include Brunei, Sabah (then 'North Borneo') and (the then British crown colony of) Sarawak – all on the island of Borneo – and Singapore, as well as the Federation (see Figure 3.1). The formal Proclamation of Malaysia was delayed by objections from Indonesia and the Philippines, but was eventually made on 16 September 1963. The new nation-state initially included all of the grand alliance territories with the exception of Brunei.[3] However, predominantly-Chinese Singapore was subsequently expelled from Malaysia in August 1965 following opposition from the People's Action Party (PAP) to the terms of the constitution. The inter-ethnic contest which led to this geo-political division was symptomatic of a broader sense and language of communalism through which the making of post-independence Malaysia(ns) has been addressed.

The formalisation of British rule in the nineteenth century had fomented and responded to existing human diversity in the Peninsula and Borneo. Various local and regional groups within and beyond the Peninsula recognised a broad *Melayu* ('Malay') heritage with roots in the Islamic Melaka kingdom. The distinctions between these and other groups, both migrants (such as the Minangkabau, Bugis and Javanese) and even, in some cases, 'indigenous' communities (now collectively known as *Orang Asli*, 'original people') were frequently blurred (Milner, 1994). Early Chinese communities, in Malacca for example, intermarried with Malays forming mixed 'Baba' groups, while later migrants working in mining or agriculture often came to form their own settlements and enclaves (Andaya and Andaya, 2001). As the Malay Peninsula was increasingly plugged into British industrial development, so labour demands and opportunities fuelled unprecedented levels of migration particularly from China (Nash, 1989) and southern India (Parmer, 1960; Ramasamy, 1994). The establishment of communally-specific institutions – from a Department of Chinese Affairs to government-approved toddy shops for Indians – formalised authoritatively-defined 'racial' identities as a feature of colonial governance. Even apparently 'innocent' bureaucratic practices, such as census-taking, devised and consolidated differences through which colonial society was ordered (Hirschman, 1986; Milner, 1994). The space for 'race' on colonial paperwork related to a specific communal place in society (Shamsul, 1997).

Colonial social geographies of place and belonging demand some further attention. The colonial economy promoted a spatial divide which mutually reinforced ethnic or racial distinctions. For the most part, Malays lived in rural *kampungs* ('villages') and on Reservation land; Indians, the majority of whom were Tamils, worked on the British plantation estates; and the Chinese on smaller plantations and in the mines. Given that a number of tin mining areas, including Kuala Lumpur, subsequently became nodes for urban development (Gullick, 1955), this colonial spatial division of labour later manifested itself as one of broadly Chinese urban, versus Malay and Indian rural, inhabitation. Maureen Sioh draws a further contrast between the mobility of the British and other Europeans, on the one hand, and the rest of the population[4] who knew their respective places, on the other (Sioh, 1998).

The non-European 'races' were not all equal in the eyes of the colonial bureaucracy. The 'special position' of the Malay rulers and their Malay subjects was 'adverted to time and again by the British administration' (Comber, 1983: 11) and the privileged rank of the Malay elite was recognised by the creation of an administrative stratum second only to the British colonial service (Khasnor, 1984). Colonial administration was carried out by an appropriation and extension of the traditional Malay Sultanates.[5] British policy thus both presumed and conferred legitimacy upon conceptions of the Malays as the rightful inhabitants of the Peninsula territories. The Malay Reservation Act, which provided land that could not be sold to 'non-Malays', is evidence of British colonial understandings of the 'indigenous' Malay whose position was under threat from immigrant others. The Malay was imaginatively constructed as unsuited to or even incapable of wage labour in either tin mines or in the nascent plantation network. Other indigenous societies of Malaya – themselves (sub-)divided and classified by colonial anthropologists and administrators, but collectively referred to as 'aborigines' (see Nicholas, 2002) – were similarly allocated Reservation land.

While the broad social and spatial contours of modern Malaysia were thus identifiable by the end of the nineteenth century, it is important to emphasise that the *raison d'être* for British 'influence' and later colonisation was commercial. The most tangible contemporary sign of the colonial economy was dramatic landscape transformation both from the cultivation of new crops for sale and the clearance of previously uncultivated land for plantation estates. However, three less visible long-term geographical implications are also worthy of note. The first relates to the *external orientation of economic development* in the Peninsula and, to a lesser extent, Borneo. Economic development oriented to securing raw materials to fuel the industrial growth of Britain gave the transport network a characteristic pattern (Leinbach, 1974). Road and rail routes were forged to service the tin mines of the west coast of the Peninsula with Kuala Lumpur occupying a 'pivotal position' following the completion of the Singapore-Butterworth railway in 1909 (Lim, 1978: 96). The relatively poorly-served east coast of Peninsula Malaysia and East Malaysian states to this day reflect and contribute to a double-layered east–west disparity in the regional geography of national development (see Fauza, 1994).

A second implication concerns the *nature of commercial organisation*. Apart from the overwhelming dominance of key sectors at independence –

in particular, plantations and mining – the scale of commercial development was highly significant. Large-scale commercial agriculture in colonial Malaya paved the way for corporate-style organisation which continues to dominate agriculture and other sectors of the economy today.[6] Third is a sense of the colonial economy as an object lesson in the *transformative power of science and technology*. While colonial administrators arrogated to themselves the responsibility of ordering 'unruly' natures and cultures for the good of all, a more enduring 'rhetoric of modernisation' has legitimised capitalist expansion (Sioh, 1998: 147). However, despite growing political awareness by the 1920s, it appears that the social implications of colonial economy, including the communal division, were only brought home with the outbreak of the Pacific War in 1941 (Watson, 1996).

The Japanese, who overran British Malaya in December 1941 employed a far more rigid divide and rule policy, focusing the attention of their violence on the Chinese, on account of the Sino-Japanese conflict (which had started in 1937), and so exacerbating existing inter-ethnic tensions. In addition, whereas the British had conceived of the large Chinese minority as temporary residents, the unstable, war-torn situation in mainland China can only have added to any citizenship ambitions which had previously arisen from economic diversification (Nash, 1989). Given the strongly anti-Japanese stance of the Chinese during the war, it was perhaps not surprising that the British proposal for a Malayan Union gave non-Malays equal citizenship rights with Malays. Official national histories have it that the scheme was strongly opposed by Malays – on the grounds that it involved centralisation which would reduce the authority of the Malay rulers (who themselves initially acquiesced to the proposals) and leave the Malays as a minority in their 'own land' – and that this opposition was led by the United Malays National Organisation (UMNO). UMNO leader, Dato' Onn Jaafar warned the Malay rulers (*raja*) that in accepting the Malayan Union proposals, they had betrayed their subjects (*rakyat*). The rulers changed their mind and thus, 'the ideological position of the Malay *raja* was considerably restored' (Muhammad Ikmal, 1996: 51). A key post-colonial contradiction emerges here. On the one hand, Malay nationalism rejected and reworked British visions of a post-independence Malay(si)a. On the other hand, however, the very grounds for this opposition drew upon colonial scientific racism as the basis for communal categories and even conceptions of Malay indigenous rights.

On 1 February 1948, the British signed an agreement acceptable to UMNO to establish *Persekutuan Tanah Melayu* ('The Federation of Malaya'). This formally instituted what has been referred to as the 'ground rule' of the political system: 'the special position of the Malays and the legitimate interests of the other communities' (M. Ong, 1990).[7] The nine sultans of the Malay states were organised into a Conference of Rulers, and every five years they would elect from among themselves the *Yang di Pertuan Agong* ('King'). Recognition of the Malay sultans as sovereign monarchs meant *ipso facto* that the Federation of Malaya was a Malay state (Muhammad Ikmal, 1996).[8] Non-Malays had citizenship rights, but this citizenship did not amount to a nationality (Harper, 1996). The Malayan Indian Congress (MIC) and the Malayan Chinese Association (MCA) were established to represent the 'legitimate interests' of the Indian and Chinese communities respectively.

The communal subjects on which this political division was built did not, of course, emerge fully-formed from colonial society. Apart from the diversity of 'immigrant' groups labelled Indian or Chinese, the economic geography of colonial Malaya meant that 'Malays' also encompassed a wide range of rates of transformation and development (Sullivan, 1985). Malayness, like other social identities was (and, as noted at the beginning of this chapter, remains) in a state of construction and (re)definition (Kahn, 1992). The same may be said for Malay nationalism. After the war, *Partai Kebangsaan Melayu*, (PKMM, 'Malay Nationalist Party') promoted the idea of *Melayu Raya* ('Greater Malay nation') based on *Bangsa Melayu* (Malay nationality) which would have included much of present-day Indonesia (Muhammad Ikmal, 1992). C. W. Watson notes a tradition within the Malay Left to trace Malay nationalism even further back to the *Kesatuan Melayu Muda* (KMM, 'Union of Malay Youth') thereby denying the aristocratic origins of Malay nationalism associated with UMNO (Watson, 1996).

Non-Malay opposition was likewise not confined to the 'legitimate interests' and neat communal political 'containers' of MIC and MCA. From 1948, the Malayan Communist Party (MCP) was involved in violent conflict with the British which became known as the 'Emergency'.[9] MCP grew out of the Kuomintang and the Nanyang Communist Party 'both of which were oriented to China and matters Chinese' (Muhammad Ikmal, 1992: 265) and, after failure to mobilise Malay peasants and Indian workers with the promise of equality of rights among all nationalities, MCP was gradually compelled to promote 'Chinese rights' to retain support as it was forced into the jungle. There are three points to be made from the Emergency itself which are relevant to post-independence transformation. The first concerns the site of the conflict, the 'forest' or 'jungle' which came to signify not only unruliness, but also danger, threat and conflict. Sioh (1998) relates this to a relatively low level of concern at the destruction of rainforest in contemporary Malaysia even among supposedly 'environmentally-conscious' middle classes. A second point concerns the political measures instituted by British authorities. In particular, the Internal Security Act (ISA) which sanctioned detention without trial for indefinite periods has been used to clamp down on an increasingly wide range of political opposition and dissent (Syed Husin, 1996; Khoo, 2002). A third point relates to unintended spatial consequences of British responses and strategies which included the forced resettlement of rural Chinese into fortified 'New Villages' so as to cut off supplies to the communists in the forests (Jackson, 1991). The plan had the desired effect in security terms but also served to augment already disproportionately high rates of urban settlement among the Malaysian Chinese community (Mohd Razali, 1993).

By the time the Emergency began to peter out in the late 1950s, communal politics had been formalised. In October 1954, UMNO, MIC and MCA formed a national Alliance to represent the interests of the three main ethnic communities in Malaya. The Alliance won all but one of the 52 seats at the first national elections in 1955, following which Tunku Abdul Rahman became Prime Minister. Represented by the three communally-based political parties, independent Malaya was imagined as an always-already communally-divided territory. From the outset, the making of Malaysia(ns) was beset with contradictions which arose from seeking to graft European conceptions of an ethnically homogeneous nation-state onto a society of multiple politically-formalised communities (Harper, 1996). Debates over the formation of a Grand Alliance in the 1950s and 1960s brought to the fore another division based around the classification '*bumiputera*' which included Malays and other supposedly indigenous 'sons of the soil', particularly in Sabah and Sarawak.

Presumptions of competing communal interests helped to shape the first phase of economic development in independent Malay(si)a and extend certain (neo-)colonial trajectories. In the first place, apart from assurances at independence that there would be no nationalisation of British interests, it has been suggested that UMNO favoured a large foreign stake in the economy so as to limit the expansion of Chinese capital (Gomez and Jomo, 1997). It was British investors in particular who took advantage of infrastructure and credit facilities as well as tariff protection which was introduced to encourage import-substituting industrialisation (ISI). The promotion of export-oriented industrialisation (EOI), following the exhaustion of ISI in the mid-1960s, reduced the opportunities for domestic capitalists even further (*ibid*. 1997). This amounted to an effective maintenance of British economic control despite conceding political control. It is also important to point out, however, that the post-colonial formation was not merely reducible to an historical extension of British economic and cultural power. 'Development' recommendations by new supranational institutions espoused economic openness and limited government intervention. It was in part through these supranational economic prescriptions that the first stage of transformation in independent – but still very much culturally and politically, as well as economically, interconnected – Malay(si)a was characterised by *laissez-faire* non-intervention.

Laissez-faire economic policy meant that what little domestic participation there was in the formal economy continued to be predominantly Chinese. Reflecting, no doubt, the locus of UMNO's political support, government initiatives focused rather on rural development: the Rural and Industrial Development Authority (RIDA) provided Malays with access to credit facilities and business training; and the Federal Land Development Authority (FELDA) distributed land for the cultivation of cash crops

(Shamsul and Lee, 1988). Despite educational and economic 'special privileges', Malay capitalism remained something of an oxymoron, fuelling political and economic fears of Malays being overwhelmed in their 'own country' (Comber, 1983). The issue of advancing Malay business interests, therefore, gained momentum in the 1960s, particularly following Bumiputera Economic Congresses in 1965 and 1968 (Jomo, 1995). Frustrated Malay economic expectations reduced support for UMNO, while government efforts to enhance the Malay economic role only exacerbated Chinese dissatisfaction with MCA (Gomez and Jomo, 1997). This situation, combined with reductions in the real income levels of the poorest sections of society, contributed to the low level of support for the Alliance in the election of 10 May 1969 and ethnic riots which began in Kuala Lumpur, three days later (Comber, 1983).[10]

On 11 and 12 May, opposition parties held victory parades in Kuala Lumpur which passed through Kampung Baru, the largest Malay residential area in the capital (Comber, 1983).[11] A counter-demonstration by UMNO on the evening of 13 May sparked off disturbances between Chinese and Malays. Soon after, rioting broke out in several parts of Kuala Lumpur and a curfew was declared at 8 pm. During the week that followed, a 24-hour curfew was placed over virtually the whole of the west coast of Peninsula Malaysia – Selangor, Malacca, Perak, Kedah, Penang and Negri Sembilan (see Figure 3.1). A national emergency was declared and parliament suspended. Tunku Abdul Rahman set up a National Operations Council responsible for administration under the emergency and appointed a new cabinet. Gradually, the worst racial riot in the history of the country subsided, official figures indicating that almost 200 people had lost their lives. More than 9,000 people were arrested, of whom more than half were charged in court, and scores of vehicles and buildings were damaged. Irreparable damage was also done to the consociational Alliance which formed the first period of post-independence Malaysian development fomenting new aims of and means to making Malaysia(ns).

Bumiputeraist intervention

The riots of 'May thirteenth' shattered any sense of post-independence transformation having fostered feelings of common citizenship – united national subjects of Malaysia. Tunku Abdul Rahman set up a Department of National Unity to prepare guidelines for a national ideology and August 1970 saw the promulgation of *Rukunegara*, a set of five principles of state (Comber, 1983).[12] However, whether through lack of substance, or through the way in which it was implemented, *Rukunegara* never really caught on. It is perhaps official imaginings of the riots themselves which have haunted subsequent socio-political change. The riots 'transcended the actual event to become an ideological instrument of the state, being a powerful symbolic code for protecting Malay nationalism and curbing non-Malay assertiveness'

(Lee, 1990: 493). May thirteenth is thus said to have legitimated the exercise of (hegemonic Malay) state power to subdue (non-Malay) opposition in the name of national harmony.

Economic growth also took precedence over political principles. As Watson has put it, 'Rather than working towards the creation of a common set of political and moral values, the government in effect opted ultimately to promote the economic vision of wealth for all to the exclusion of almost everything else' (Watson, 1996: 318). In part, this meant an extension of EOI – particularly through the development of industrial zones such as Beyan Lepas in Penang (see Figure 1.1) – realised through high levels of direct foreign investment by transnational corporations (Kahn, 1996a). However, it was the projected 20-year New Economic Policy (NEP), introduced by the new Prime Minister, Tun Abdul Razak, which foregrounded a new societal division and which came to define a broad second phase of Malaysian development.

Though one of the twin objectives of NEP was the eradication of poverty for all Malaysians, irrespective of race, it is the second objective – a 'restructuring' of society to eliminate the identification of race with economic function – with which it is most frequently associated. As Gomez and Jomo suggest: 'From the outset ... the keenest interest in implementing the NEP was clearly on restructuring wealth, particularly on creating a Malay business community and achieving 30 per cent Bumiputera ownership of the corporate sector by 1990' (Gomez and Jomo, 1997: 24).[13] In the context of NEP, 'Bumiputeraism' (Brown, 1994) came to articulate a divide between those citizens eligible to a range of state interventionist positive discrimination (or 'affirmative action') on account of their indigeneity, on the one hand, and the rest of the population, on the other (Gomez and Jomo, 1997). Affirmative action came to include ethnic quotas favouring *bumiputeras* in education, employment in the government sector and private corporate enterprises, the award of government contracts and stock ownership in corporations.[14] This, along with post-1970 assertions that the national culture 'must be based on the cultures of the people indigenous to the region' (National Cultural Policy; in Tan, 1992: 283), meant that *bumiputera* became 'the ideological cornerstone of the modern Malaysian state' (Harper, 1999: 229). *Bumiputera*- and especially Malay-centredness was founded upon a much more dominant role for UMNO in the new *Barisan Nasional* (BN, 'National Front') than in the Alliance which it replaced.

NEP operated on the assumption of an unambiguous distinction between *bumiputera* and non-*bumiputera* (Milne and Mauzy, 1986). The former would be the beneficiaries of a new interventionist stage in Malaysian national development, not through a redistribution of existing wealth, but through a more equitable distribution of newly-created wealth from continued economic expansion (Gomez and Jomo, 1997).[15] However, it is the uncritical acceptance of either the assumptions and generalisations on which the distinction is premised – typically counter-posing the poor, rural

Malay with the entrepreneurial, urban Chinese – or that *bumiputera* and non-*bumiputera* are respectively undifferentiated beneficiaries and 'victims' of NEP, which is said to have contributed to an 'ethnicisation of knowledge' about Malaysia (Shamsul, 1996a: 482). On the one hand, in its early stages in particular, NEP so antagonised some Chinese that it prompted emigration by some of its more educated members (Hefner, 2001) as well as capital flight (see Jesudason, 1990). On the other hand, while the purported intention of NEP has been as a move of positive discrimination in favour of Malays *en masse*, this is not the only way in which the policy may be interpreted in practice. Many commentators suggest that it was not so much *bumiputera*, or even Malays, who were the main beneficiaries of the policy, as a Malay middle class in particular (Kahn, 1996b; see also Crouch, 1985).[16]

The demands of the small but expanding Malay middle class in the 1960s for greater government intervention were bound up with claims of economic dispossession, which were largely ignored by the Alliance government. In his book, *The Malay Dilemma*, published in 1970, Mahathir Mohamad referred to 'ridiculous assumptions', inherited from the British that 'the Malays wished only to become government servants' and so not participate in the private sector of the economy (Mahathir, 1970: 15). Mahathir, the Malay 'ultra' (Khoo, 1995)[17] who had been expelled from UMNO because of a highly-critical open letter to Tunku Abdul Rahman in June 1969, raised the problem:

> The Malay dilemma is whether they should be proud to be the poor citizens of a prosperous country or whether they should try to get at some of the riches that this country boasts of, even if it blurs the economic picture a little.
>
> (Mahathir, 1970: 61)

In keeping with rising Malay middle-class economic aspirations, this was no dilemma at all. The real issue was how to promote Malay participation in the economic life of the country through 'constructive protection' (*ibid.* 31).

The Malay Dilemma, therefore, is to be understood as part of a broader rationality of Malay development associated with a new generation of Malay socio-political authorities and out of which NEP emerged. Mahathir played no official part in formulating NEP – his political exile only ending in 1972 when he was invited to rejoin UMNO by the new Prime Minister, Tun Abdul Razak – and *The Malay Dilemma* was banned in Malaysia. However, the book provides an important window into what has been termed 'the dominant ideological framework of the post-1969 polity' (Khoo, 1995: 27). Many of the ideas raised in *The Malay Dilemma* were put into practice in the administrations of Tun Abdul Razak and his successor Tun Hussein Onn. Of particular interest to this geographical analysis are ideas on urbanisation and its role in Malay(sian) 'development'. In *The Malay Dilemma*, there is a spatial dimension to Mahathir's understanding of Malay 'exclusion from the commercial life of the country' (Mahathir, 1970: 37). Noting the 'importance of urbanisation in the progress of a community' (*ibid.* 79–80), he argues that the colonial economy condemned Malays to static, backward, rural areas. Malay 'backwardness' then is, in part, interpreted as a consequence of the geography of colonial (under-)development.

In explanations of Malay 'debilitation' in *The Malay Dilemma* which rationalise a politics of intervention, Mahathir essentially shared stereo-typical colonial prejudices against the 'lazy native' (Alatas, 1977).[18] Mahathir alludes to pre-colonial racial geographies in an environmental determinist understanding of Malay backwardness *vis-à-vis* the Chinese: 'If we want to examine the development of the Malays in Malaya we must first study the geography of Malaya and determine its effects on them' (Mahathir, 1970: 20). Malays are said to have lived historically on sparsely populated lush tropical plains which, 'were not conducive to either vigorous work or even to mental activity' and where, 'even the weakest and the least diligent were able to live in comparative comfort, to marry and procreate' (*ibid.* 21). In contrast, Chinese history is depicted as one in which 'life was one continuous struggle for survival' and so 'through four thousand years or more ... weeding out of the unfit went on' (*ibid.* 24).

In addition to these broad colonial hereditary and environmental 'explanations', the Malay *kampung* ('village') is singled out as the locus for primitive social practices and values: 'Malay partiality towards inbreeding' in the *kampung* is contrasted not only with the Chinese – whose 'custom decreed that marriage should not be within the same clan' – but also with 'town Malays' who intermarried with Indian Muslims and Arabs (*ibid.* 24).[19] Malay *adat* (or 'custom') is cast in feudal terms as an impediment to their acceptance of philosophy, sciences and other new ways of thinking. Malays are said to be characterised by: 'a combination of tolerance and *laissez-faire* attitude' (*ibid.* 139); 'natural courtesy' (p. 142) (understood as a sign of weakness in the eyes of non-Malay 'competitors'); a formality which 'does not condone innovations' nor 'encourage change and inventiveness' (p. 157); and a strong belief in the 'proper way to do things' which only serves to negate the quest for 'logical explanation of why the prescribed way is correct and acceptable' (p. 157). Indeed, as a whole, the 'Malay social code' is 'somewhat anachronistic and can only lessen the competitive abilities of the Malays and hinder their progress' (p. 171). *The Malay Dilemma*, in short, is a (post-)colonial problematisa-tion of Malayness, rendering it in need of 'development'.

Dr Mahathir casts the urban as a potential incubator of modern Malay-ness, a remedy[20] for backward values of the *kampung*:

> The fact of urbanisation alone involves a process of physical and psychological uprooting of the Malays from the traditional rural society. There can be no doubt that with this uprooting, old values and ways of life must give way to the new.
>
> (Mahathir, 1970: 113)

Thus, a planned programme of Malay urbanisation is imagined as crucial if the Malays are going to 'master' modern ways. Idioms of evolutionary biology are mobilised in the cause of a more general Malay orientation towards the new, the modern. Yet, despite clear Social Darwinist under-standings, *The Malay Dilemma* claimed to advocate not so much evolution as 'revolution'. Mahathir contrasts evolution, which 'cannot be properly controlled in speed or in objective' with revolution, which at least has the potential to be 'creative and orderly if the mechanics are understood by those best able to carry them through' (*ibid*. 103). The supposedly debilitating effects of Malay heredity and environment can, therefore, be overcome by a 'systematic and co-ordinated orientation of the Malays towards progress' (*ibid*. 113). This would appear to draw upon a neo-Lamarckian faith in the inheritance of acquired characteristics which suggests that modifications can be built up and the tempo of evolution increased (Khoo, 1995).[21] For Mahathir, this is fortunate given that, unlike the Chinese, Malays 'do not have four thousand years to play around with' (Mahathir, 1970: 31).

The Malay Dilemma thus set the tone for large-scale NEP social engineering legitimised through Bumiputeraist indigeneity, but directed largely at the Malays. Two further points follow on from this. The first concerns the spatiality of this Malay-centred regime of development. The intertwining of social and spatial wealth disparities was an important motivation for strategies to homogenise development throughout the national territory. Thus, from the *Third Malaysia Plan* (Malaysia, 1976), there were attempts to spread industrial and urban development more evenly across the country. In line with Mahathir's prescriptions, the state also promoted the migration of rural Malays to existing urban areas not least to greater Kuala Lumpur which had become the nation-state's dominant urban region industrially as well and politically (Malaysia, 1971; Chan, 1994; and see Hamzah, 1965 on the emergence of a greater Kuala Lumpur);[22] and the construction of housing estates, industrial centres, whole new towns, in new areas (Lee, 1987). Urbanisation was thus understood as one means by which Malays could break free from colonial stereotypes, which had been perpetuated by the Alliance in the first period of post-independence development, and so take their 'proper place' in Malaysian society and economy.

However, second, it is important not to accept uncritically generalised conceptions of Malays as the 'winners' of an ethnic-centred regime of state bio-political investment. While the rhetoric of NEP cast Malays as its principal beneficiaries – the supposed communal divisiveness of which is well known – this meant their undergoing 'revolutionary' reorientation towards development understood almost exclusively in quantitative economic terms (Mahathir, 1970). 'Progress' was perhaps less apparent in the lived realities of individual and collective Malay subjects. Malay women, in particular, comprised a new peri-urban proletariat, working for low wages in the virtual absence of unionisation (see, for example, Ong, 1987). Some commentators have suggested that this 'revolutionising' of Malay modes of living and working was an important domestic factor in explaining Islamic revivalism or resurgence in Malaysia.[23] According to Chandra Muzaffar, such religious trends articulated and reflected the spiritual alienation which many Malays had experienced from urbanisation and Western mores (Chandra, 1987). For some Malay Muslims, the response was anti-urban, anti-Western and anti-development.

The Challenge, a collection of essays written in the 1970s, by which time Mahathir had become a prominent cabinet member, represents a broader renewal of attempts to reconcile Islam and modernity (Mahathir, 1986).[24] In the Introduction, Mahathir laments:

> One of the saddest ironies of recent times is that Islam, the faith that once made its followers progressive and powerful, is being invoked to promote retrogression which will bring in its wake weakness and eventual collapse.
>
> (Mahathir, 1986: np)

He subsequently alludes to a 'Golden Age of Islam' (*ibid*. 23) when Muslims either founded or developed various fields of 'secular' knowledge, which is contrasted with a 'retreatist' tendency among Muslims in Malaysia that could only jeopardise the position of the Malays and Islam by exhorting believers to 'turn their backs on the world' (*ibid*. 18). Malay development is given a religious imperative since it is precisely 'the mastering and use of modern ways which can safeguard the position and security of Muslims' (*ibid*. 81). There are important connections here with earlier currents of modernist Islam including the nineteenth century Islamic reformist movement and even Malay nationalist figures such as Za'aba (Hussin, 1993). However, there was also an electoral imperative for the 'Islamicisation' of national development discourse and symbolism.[25] Although Malaysia's Islamic 'resurgence' was a movement and a set of discourses which were themselves 'characterised by disagreements, debate and fragmentation', what otherwise diverse Islamic political movements shared was that they were all in one way or another prepared to criticise UMNO on Islamic grounds (Jomo and Ahmad, 1992: 79). If Bumiputeraism represented a new meta-fault line for 'national' development, Islamic resurgence, including state Islamicisation, fomented new political divisions.

Industrial developmentalism

Mahathir's ascension to the highest office in June 1981, at the mid-point of NEP, set in motion a further reworking of aims of and means to development in Malaysia. The first non-aristocratic Malaysian Prime Minister, Mahathir marked what Gordon Means regards as a 'second generation' of Malay leader (Means, 1991). Mahathir's father was a schoolmaster – neither Malay aristocrat nor peasant – and so part of what Khoo Boo Tiek understands as an urban colonial Malay *petit bourgeoisie* (Khoo, 1995). The son of a man whose ability had enabled a degree of social mobility not usually associated with his class and ethnicity, Mahathir, the modernist Malay nationalist, would free Malays from the historical constraints of feudalism (Rehman, 1993: 191)[26] and (post-)colonial racial stereotyping. More than this, Mahathir, the Malaysian Prime Minister, would seek to chart an autonomous course of national development – one not conceived as mimicking Western experience. In what I consider here to be a third broad historical stage of post-independence national development, Mahathir looked to build Malaysia industrially by learning from the developmental examples of rapidly-modernising East Asian states.

The impressive economic performance not only of Japan, but also South Korea, followed, in turn, by other so-called 'Newly Industrialising Countries' (NICs), Taiwan, Hong Kong and Singapore in the 1970s, undoubtedly influenced the decision of the Malaysian Prime Minister to look 'East' for new models of economic development. The Look East Policy (LEP), announced towards the end of 1981, was one in which Mahathir strove to teach Malays(ians) the economic ways of Japan and South Korea. Some Japanese firms, in particular, were already well-established in Malaysia and, in line with the ethnic quotas demanded by the 1974 Industrial Coordination Act, contributing to the nurturing of a Malay class of managers and professionals (Smith, 1999). LEP's impact was perhaps most marked in relation to the role of the state: the state-owned Heavy Industries Corporation of Malaysia (HICOM) was brought directly under the control of the Prime Minister's Department to promote heavy industrialisation following South Korean principles (Drabble, 2000); and 'Malaysia Incorporated', which helped to bridge (what Mahathir saw as) the 'ridiculous' division between the Malay state and the non-Malay private sector, drew directly upon the Japan Incorporated concept. An extended state role in industrialisation also suggested further possibilities for the decentralisation of regional development beyond the existing Peninsula west coast urban centres and especially Kuala Lumpur-Klang Valley (Ruland, 1992).

However, perhaps the principal attraction of looking East was in bringing into view a suitable value system for Malaysian, and especially Malay, economic development. This may be understood in two ways. First, what was so attractive about Japan, in particular, was the perception that its work ethic was a 'cultivated value system' (Mahathir, 1985: 155). This suggested a hope and belief that a similar cultural system could be propagated for economic 'success' among Malays. Second, both Japan and South Korea were free from suggestions of Sinocisation. A problem with *The Malay Dilemma* had been that Malay rehabilitation could appear to be a process of making Malays more like the 'competitive' and 'resilient' non-Malay, and especially Chinese, Malaysians (Mahathir, 1970). There were (and still are) obvious political difficulties with a situation in which the foremost Malay nationalist seemed intent not only on making the Malays less Malay, but – and there is a difference in the Malaysian context – on actually making them more 'non-Malay'.

The new direction for cultural-economic policy meant not only looking towards a rising East. It also meant a turn away from, a conscious rejection of, the West:

> We see the Japanese have made headway while the West has not only not made headway, but appears to be regressing. So in order for Malaysia to progress, we have to learn from the better example and the better example is the Japanese example. That is why we now want to look East where before we were looking West.
>
> (Mahathir, cited in Das, 1982: 38)

Already in *The Challenge*, Mahathir had caricatured a 'West' suffering a 'perversion of values': from striking workers, demonstrating students and disrespectful youths to drug taking, homosexual marriages and pornography (Mahathir, 1986: 91). While 'Eastern-inspired' industrial developmentalism in Malaysia brought about its own moral panics – perhaps most notably concerning female factory workers newly liberated from *kampung* space and values (Jamilah, 1994) – the imaginative reconstruction and othering of the West was an important rhetorical device for nation-building. The issue of values (in)appropriate to the East emphasised commonalities rather than differences between Malaysians (Khoo, 1995: 42).[27] There is arguably a conceptual link between discourses of Asia(n regional identity) and the construction of a pan-ethnic Malaysian national identity. The growing assertion of 'Asian-ness' has been particularly prominent in foreign policy activities through participation in regional fora (Camroux, 1994),[28] the most successful of which has been the Association of Southeast Asian Nations (ASEAN). As David Camroux points out, 'as a multi-ethnic body, it symbolises for the Malaysian leadership a model to tacitly proffer to Malaysian society as a whole' (*ibid.*, 1994: 19).

As we have seen earlier, conceptions of an 'Asian way' and its novelty are deserving of critical scrutiny. Certainly in the 1980s and 1990s, there was a growing conception of a *new* 'Asian model' based on the developmentalist strategies which had fostered rapid economic expansion in Japan and subsequent 'waves' of Asian countries.[29] Aside from the considerable differences between the development(al) practices of each of these countries, however, there was perhaps less that was new in the discourse of 'Asia' than was newly emphasised. Descriptions of the importance of moral values, order and discipline, for example, echoed themes of colonial conservatism (Khoo, 1995).[30] And Mahathir's understanding of Western regression in civilisational terms (Mahathir, 1986)[31] meant that, 'Changes to the global economic environment seemed to fit in logically with his Social Darwinism now transposed onto the world of nation states' (Khoo, 1995: 66). Theories and conceptualisations from the 'Western' colonial past thus remained very much part of the post-colonial Asian present (and futures).

Conservative economic ideas, of course, were difficult to reconcile with NEP state interventionism. This tension came to the fore during the economic down-turn of the mid-1980s (Jomo, 1995).[32] While Malay wealth and commercial participation had increased in quantitative terms through NEP affirmative action, had this been accompanied by new cultural attitudes to money, land and property (*The Malay Dilemma*)? And while expanded commercial and industrial opportunities had realised Malay middle classes and even workers as consumers (Smith, 1999), had new work roles also served to instil greater discipline and work ethics (*The Challenge*)? It has been noted that, following the very evolutionary biological principles which Mahathir himself espoused in *The Malay Dilemma*, there was perhaps no reason why such outcomes should ever have been expected. NEP was, after all, an 'attempt to shield the *Bumiputeras* from the ravages of commercial risk' (Adam and Cavendish, 1995: 14). The series of public sector cut-backs announced during economic downturn in 1986 exposed Malay dependence on state licences and contracts. NEP, a policy ostensibly oriented towards a 'more equitable' distribution of newly-created wealth, was redundant without economic growth and it was in this context that Mahathir temporarily (and quietly) suspended NEP *bumiputera* equity requirements (Khoo, 1995: 140).[33]

This temporary measure may in retrospect be seen as diagnostic of a more deep-seated governmental dilemma. While Alasdair Bowie understood the Mahathir era – and the Heavy Industries Policy, in particular – in terms of an increasing state economic role in the pursuit of NEP ethnic redistribution targets (Bowie, 1991),[34] subsequent commentators considered recession in the mid-1980s as having provided the opportunity to implement long-held conservative Mahathirist economic ideas (Khoo, 1995: 143).[35] In *The Malay Dilemma*, Mahathir had characterised a healthy society and economy as one in which there is 'free enterprise' and economic competition between individuals and groups (Mahathir, 1970: 52).[36] And Mahathir came to power expressing a belief that Malays should *berdikari*, an abbreviation for *berdiri kaki sendiri* ('stand on their own two feet') (Das, 1981: 30). More tangibly, Mahathir's appointment as Prime Minister coincided with a halt to the creation of State Owned Enterprises (SOEs) other than those for heavy industry and, in 1983, he announced a privatisation policy. According to Jomo K. S., therefore, the apparently 'dramatic reversal of preceding Malaysian government policy … was very much consistent with his [Mahathir's] own personal ideological and policy preferences' (Jomo, 1995: 2).

I wish to re-emphasise that the ongoing reworking of Mahathirist rationalities of government occurred in relation to events and idea(l)s elsewhere, including in 'the West'. The evolution of Mahathirist policy preferences, no less than earlier stages of development, was not taking place in an international vacuum. Following the collapse of oil prices in early 1986, there was considerable pressure on Malaysia to respond with policy changes favoured by the Bretton Woods institutions (Jomo, 1995). More broadly, the 1980s saw the pervasive influence of the market conservative ideology of the 'new right' – Thatcherism and Reaganomics.[37] The world which Thatcherism, in particular, described fitted in well with Mahathir's understanding of continued Malay debilitation. While Thatcherism suggested that the state's assumption of responsibility for ensuring welfare and employment had led to 'a decline in individual initiative and responsibility' (MacInnes, 1987: 49), Mahathir believed that NEP state expansion in Malaysia had perpetuated an existing dearth of Malay enterprise. And, like Thatcher, Mahathir mobilised patriotic sentiments to gain support for the project of national renewal (Khoo, 1995).[38] Contemporary as well as colonial manifestations of conservative ideology, therefore, shaped Mahathirist economic convictions.

These convictions were not shared by all members of UMNO. Indeed, it may be contended that the split in the party which occurred in 1987, while certainly motivated, in part, by individual lust for power and greater access to government licences and concessions (Crouch, 1992), also reflected conflicting economic policy preferences. Khoo Kay Jin (1992) argues that the division can be interpreted in terms of policy differences between two broad UMNO factions: on the one hand, the Musa-Razaleigh camp[39] was concerned with the implementation of NEP and how best to achieve existing *bumiputera* ownership targets; the Mahathir camp,[40] on the other hand, was convinced of the need to completely overhaul or even to replace NEP. As far as Mahathir was concerned, many Malays, afflicted by a 'subsidy mentality', had come to 'think that NEP means a free gift', and so would continue to rely on the state as a 'crutch' (cited in Kahn, 1996b: 17). Malay modernisation in Mahathirist terms would henceforth mean more than just a series of numerical targets as in NEP; it would mean the formation of entrepreneurial, hard-working, robust Malays – a Malay capitalist class (Khoo, 1995). The ultimate victory of the Mahathir camp, therefore, may be said to have signalled possibilities for a fourth (post-NEP) period of post-independence national development. For Khoo Kay Jin, however, it also meant 'an administered and increasingly authoritarian society and polity under a hypertrophied executive with greatly reduced space for popular participation except as a disciplined workforce' (Khoo, 1992: 71).

The so-called Operation Lallang in 1987, which saw the arrest of some 106 citizens under the Internal Security Act (ISA) following an apparent resurgence of the 'ethnic problem', marked a shift to greater authoritarianism in Malaysia. In concluding this section, I raise two issues in relation to Operation Lallang. First, since NEP state interventionism had been justified in terms of overcoming the ethnic problems which were believed to have fuelled the events of 13 May 1969, the continued perception of ethnic difference as potentially dangerous suggested the failure of NEP. A new development policy might, therefore, be expected to de-emphasise the 'Bumiputeraist' divide instituted by NEP. As shown above, a reduction in state interventionism also fitted in with Mahathir's personal economic ideas. And, it might be suggested that accelerating processes of economic globalisation and regionalisation, which valorised non-Malay transnational cultural links, provided impetus to a move away from a Malay- or *bumiputera*-centred national policy. However, as Jomo pointed out, any move away from NEP would be threatening to a Malay community 'led to believe that all the gains they have made since 1970 have been due to the NEP' (Jomo, 1994: 4).

The second point relates to a rather different reading of Operation Lallang. Loh and Kahn highlight the fact that, 'the majority of those arrested were government critics and on a number of issues, government ministers and members of the ruling coalition had been responsible for playing up and aggravating the tensions' (Loh and Kahn, 1992: 3). Government critics included not only members of religious and opposition political organisations, but also individuals involved in human rights, environmental and other civil society issues. The point here is not one of downplaying the social and political significance of ethnic difference, but rather to emphasise that this had become increasingly inadequate in accounting for opposition to state development policy and practice. Such a view is corroborated in part by a recognition of growing critique from sections of the Malay community, the supposed beneficiaries of NEP. Apart from Islamic resurgence noted above, 'Malay traditionalism' has been identified among the urban middle classes – the very social stratum carved out by affirmative action opportunities – as a form of resistance to social changes and the erosion of identity which have accompanied modernisation (Kahn, 1992). Malays have also formed part of movements opposing increased authoritarianism (Saravanamuttu, 1989). As the twenty year course of NEP drew to an end, therefore, speculation and anxiety over what would or should replace it was bound up not only with issues of ethnic division, but also with newer debates ranging from the environment to culture to human and political rights.

The way forward?

The debate and speculation over what would replace NEP came to an end on 28 February 1991 when Prime Minister Mahathir delivered a speech entitled 'Malaysia: The Way Forward' to the Malaysian Business Council. The contents of this paper have subsequently been officially promoted and popularised as *Wawasan 2020* ('Vision 2020') – to enable Malaysia to become a 'fully developed country' by that year (Mahathir, 1993: 404).[41]

While the National Development Policy (NPD), announced later in 1991, was the official successor to the NEP, it is Vision 2020 which has come to frame broad parameters for a post-NEP society. The political economic context in which Vision 2020 was announced is significant. Rapid economic growth after 1986 not only appeared to vindicate Mahathir's policies; it also helped to lure back many of the UMNO dissidents who had joined Razaleigh following his defeat in the leadership contest. This undermined the multi-ethnic counter-coalition which challenged *Barisan Nasional* at the general election in October 1990 and so, despite some impressive gains by the Islamic Party (PAS) including winning control of Kelantan (Crouch, 1992), *BN* retained its two-thirds majority in parliament. Khoo Boo Tiek thus understands Vision 2020 as a Mahathirist ideological 'summing up' made possible by the consolidation of the Prime Minister's political position after the 1990 election (Khoo, 1995).

Certainly, the substance of Vision 2020 was largely familiar. The need for an end to 'the identification of race with economic function, and the identification of economic backwardness with race' (Mahathir, 1993: 405) was reminiscent of NEP in both content and phraseology. In keeping with more recent ideas, Mahathir emphasised that constructive protection alone will not yield 'an economically resilient and fully competitive Bumiputera community' (p. 407). Rather, the private sector is to be the 'primary engine of growth' (p. 419), the 'cornerstone of our national development and national efficiency strategy' (p. 410). Calls for an economy which would be 'more lean, more resourceful, more productive and generally more competitive' (p. 413), stressing the importance of 'work attitudes and discipline' (p. 415), could have come straight from Look East. From the Islamisation policy, there was the reminder that development 'cannot mean material and economic advancement only' (p. 406). In reasserting a need for 'an Accelerated Industrialisation Drive', (p. 414) 'economic liberalisation' and the 'freeing of enterprise' (p. 411) too, Vision 2020 was advocating 'more of the same'.

However, I am concerned with Vision 2020 here less as a culmination of Mahathirism than in terms of an on-going shift to a broad fourth stage of post-independence Malaysian development. Aside from Mahathir's changing personal policy preferences, Vision 2020 was both a response to and appropriation of various forms of critique of previous development, especially in the late 1980s. In contrast to the ecological and environmental destruction which was perceived to have resulted from 'development' in the past, for example, Vision 2020 stressed that 'the beauty of our land must not be desecrated – for its own sake and for our economic advancement' (Mahathir, 1993: 418). Promising a 'qualitatively superior' society, 'The Way Forward', outlined a would-be hegemonic political vision, incorporating definitions of development which emerged as contest to (what was now cast as) an old period of modernisation. In line with notions of the constitutive role of contest discussed in the previous chapter, therefore, I emphasise not closure (or Mahathirist 'summation'), but rather the dynamic nature of Vision 2020 as part of broader political rationalities. What is really interesting, then, is how Vision 2020 is bound up with wider political and socio-cultural contest over post-NEP Malaysia and strategies for its realisation.

I raise three points in relation to the reworking and implementation of ideals of Malaysian socio-economic transformation. The first concerns the apparently unambiguous optimism of state development discourse. Despite the fact that 'development' is a much-maligned concept after successive decades of 'failure' in many countries of the world (Escobar, 1995), Vision 2020 exudes a modernist, developmental faith in the possibility of realising planned national futures. This is perhaps, in part, because Vision 2020 development is understood as self-prescribed and self-administered and is thereby untainted by suggestions of the exercise of Western power (see Crush, 1995). As we have seen in the previous two chapters, the discursive emergence of Malaysia as part of a New Asia/East led scholars of Southeast Asia to speak not only of development in this imagined region, but also of non-Western modernities (see also Wee, 1996; Ong, 1996). Vision 2020 exuded a confidence of being able to attain 'developed status'[42] through a nationally-specific way forward (Mahathir, 1993: 408).

This leads on to a second point which concerns attempts to articulate distinctively 'Malaysian' conceptions of development. In part, state discourse in the 1990s represented an extension of ideas from the Look East policy which rejected the non-East as a suitable or desirable model for Malaysian socio-economic transformation. In LEP, 'Asian-ness' was mobilised as a set of values and practices to revolutionise Malaysian, and especially Malay, practices. However, in the 1990s, the culturalist discourse of Asia was increasingly accompanied by a valorisation of putatively Malay(sian) values. This is evidenced from a transcoding of the concept of *kampung*. Whereas in *The Malay Dilemma*, '*kampung*' was synonymous with that community's supposed genetic and cultural debilitation (Mahathir, 1970), *kampung* now increasingly forms the basis of Malay(sian) urban design and lifestyle ideas, often in contradistinction to Western modernist development (Kahn, 1992, 1994; Bunnell, 2002c; Goh, 2002a and 2002b). *Kampung* is thus incorporated into the construction of new aims, means and markers of national development which obscure the constitutive externalities of hybrid(ising) modernities.[43]

A third point concerns the widely-held belief in Mahathir's personal dominance over the development planning process. In the mid-1990s, Mahathir was not only the longest serving Prime Minister of the country – one in whose hands was concentrated a perhaps unprecedented level of personal political power – but was popularly regarded, both within Malaysia and internationally, as *the* man behind Vision 2020. Mahathir became the 'architect of developed Malaysia' (*Asian Editor*, 1997: 29), the 'master planner' who was 'rebuilding Malaysia his own way' (*Time*, 1996:

front cover). Such nation-building, of course, had intertwined material and cultural or psychological dimensions. The former referred, in part, to a series of high-profile national infrastructure including the 869 km North-South Highway, completed in 1994, which spans the entire west coast of Peninsula Malaysia from the Thai border in the north to Johor Baru in the south (Naidu, 1995; refer to Figure 3.1). But Mahathir's imaging in relation to these projects was also significant: rarely a suit and tie-clad official, he was more commonly depicted in boiler suit or baseball cap; not 'above' development, but a hands-on participant in the work of national (re)construction. The iconography of development projects – built by Malaysians for Malaysians – connects with conceptions of a national *révolution mentale*. National landscapes were understood as performing something, bolstering self-confidence as part of Mahathir's 'relentless crusade to change the Malaysian mentality and mindset' (*Asian Editor*, 1997: 38), to show the nation that *Malaysia boleh* ('Malaysia can').

An understanding of nation-building in terms of revolutionising the outlook of citizens and enlarging their capacities relates back to ideas of governmentality discussed in the previous chapter. The contested (re)definition of the aims and objectives of Malaysian development cannot be finally disentangled from politico-cultural strategies to realise them. Popular acceptance of the characteristics of, and means to, the ideal society outlined in Vision 2020 may be expected to promote similar systems of evaluation in citizens. Vision 2020 efforts to represent the norms, values and aspirations of the society as a whole, therefore, are bound up with issues of individual and collective conduct as much as with political legitimacy. However, the latter is also significant. Shamsul A. B., for example, noted how Vision 2020 'has developed into a kind of popular public idiom' and how the word *Wawasan* ('Vision') and the figure '2020' were used in everyday situations ranging from the advertising of cinemas to the names of cakes (Shamsul, 1996b: 327). As a popular political slogan rather than an intellectual theory, Vision 2020 had, by the mid-1990s achieved a very wide circulation.

All this is not to suggest that state ways of seeing were wholeheartedly endorsed or passively accepted by Malaysian society as a whole. Apart from everyday contest to would-be hegemonic discourse, the policies and practices of Malaysia's political economy came under increasingly critical scrutiny. Jomo K. S. argued that any benefits which might have resulted from a purported shift in priorities from redistribution to growth were being undermined by the award of privatisation contracts to politically-connected individuals, predominantly Malays: 'the privatisation program has actually accelerated private Malay accumulation, arguably at the expense of the community as a whole, and certainly its poorer members' (Jomo, 1995: 6). Moreover, failure to award contracts by competitive tender contradicted notions of efficiency-through-competition used to valorise privatisation policies. In this way, public sector cut backs and privatisation only served to extend perceptions of political economic impropriety. In particular, the executive was understood as possessing unprecedented power to select 'winners' – individuals who are able to buy state assets at 'knock-down' prices (Jayasankaran, 1995a: 30). But this forms part of a more general intertwining of business and politics, often referred to as 'money politics' (Gomez and Jomo, 1997). In addition, Tim Harper identified a critical Malaysian discourse of 'accountability', which came to the fore as a response to Mahathir's authoritarian moves in the late 1980s, but which continued to articulate expectations on the use of, and limits to, political power in the 1990s (Harper, 1996: 251). This may be said to encompass not only suggestions of high-level political misuse of funds, but also more general dissatisfaction with the 'top-down' operation of Malaysian development.

Nonetheless, *Barisan Nasional's* overwhelming victory at the general election in 1995 was a massive endorsement of Mahathir's post-NEP vision. The electoral landslide evoked parallels with the first Alliance victory on the eve of independence and reflected similar optimism and the promise of future prosperity (Harper, 1996: 251). During the election campaign, the opposition parties were cast as either religious fanatics (PAS) or racial extremists who championed the cause of a single community (Democratic Action Party, DAP) (Liak, 1996). *BN*'s success, therefore, not only 'lent unprecedented credibility to its claim to be the party of all Malaysians' (*ibid.* 217), but suggested an end to ethnic politics and the possibility of a new kind of society. An international dialogue on Islam and Confucianism in March 1995 initiated by the Deputy Prime Minister, Anwar Ibrahim, exemplified a broader shift away from the Malay-centred national cultural policy. A growing vogue of multiculturalism, at least rhetorically, enabled acceptance of and even support for, the subordination of ethnic politics to expanded conceptions of national growth and development – the on-going birth of a post-NEP society.

However, while the triumphalism which surrounded the election dealt a severe blow to political opposition, the road was by no means clear to simply replace NEP state interventionism with 'free competition'; or the 'Bumiputeraist' divide with a multi-cultural national model. For one, the pervasive belief that Malays would be left behind without affirmative action persists. Ironically, Mahathir was perhaps increasingly contending with the very ethnic regime of representation to which he had contributed. Optimistic descriptions of universal benefits to be derived from a united nation were intended to erode the logic of inter-ethnic competition (Shamsul, 1996b: 331).[44] Mention of '*Bangsa Malaysia*' in 'The Way Forward', in particular, was highly significant: a Malay leader mooting the possibility of a 'level playing field' arguably functioned as a sign for a Malaysia beyond constructive protection (Hiebert, 1996). Apart from any constitutional issues which challenge the current basis of the state, however, *Bangsa Malaysia* collides with deep-rooted communal identifications. So heated was the opposition to the impending end of NEP benefits that Mahathir was arguably forced to respond with the term *Melayu baru* ('new Malay') expressing his optimism that Malays could now compete successfully with other groups both within Malaysia and globally (Khoo, 1995).

The global competitiveness of so-called *Melayu baru* here connects with a broader theme of the importance of 'out there' for Malaysia's 'in here' that runs through this chapter (see also Mee, 2002: 70). Just as constitutive externalities shaped historical geographies of modernity, so reworkings of division and development in 1990s Malaysia were bound up with globalisation, regionalisation and transnationalism. These processes and imaginings had, and continue to have, significant implications which I summarise into four key points.

1 Multicultural rescripting of national identity

Economic globalisation and regionalisation are bound up with potentially radical reworkings of Malay and Malaysian nationalism. The Bumiputeraist economic privileging or protection of the main ethnic community sits uneasily alongside imperatives of economic liberalisation associated with Malaysia's global shift. Export Processing Zones (EPZs) established in Malaysia since the early 1970s (Rasiah, 1995) provided global spaces of exemption from ethnic employment quotas as well as from a range of other national development regulations. More recently, the apparently growing importance of regional cultural-economic connections in 'networked' forms of economy (see Castells, 1996) has prompted a state rescripting of national identity. Malaysia's transformation from exporter of commodities to producer of manufactured goods is the result of investment not only from Western and Japanese MNCs, but also from second-tier 'tigers' such as Taiwan, Hong Kong and Singapore. According to David Camroux (1994: 8), for the latter investors, 'Malaysia's attraction lay not only in its relatively cheap labour force but also in the existence of a class of Chinese-Malaysian entrepreneurs and executives'. More than this:

> The concentration of multinational corporations in Penang and the federal territory of Kuala Lumpur – two areas in which the Chinese-Malaysian are in a majority – is not accidental. Coming from the same Hokkien and other clan groups as found in Taiwan and Singapore, these Chinese-Malaysian entrepreneurs provide a relay for second-wave NIC investment in Malaysia. As a consequence the pressures coming from globalisation were running, and continue to run, counter to a narrow defence of Malay interests.
>
> (Camroux, 1994: 8–9)

For Camroux, then, the increasing multiculturalism evident in official representations of the nation since the 1990s – such as the multi-racial iconography used in launching the 'national car', the Proton Saga – was more than merely coincidental. Rather, a perception of the economic importance of non-Malay transnational connections has served to unsettle Malay-centred constructions of national identity.

Political tensions arise here from the transnational reconfiguration of Malay nationalism itself. On the one hand, in the context of a valorisation of extra-national connectivity, the 'problem' of Malayness becomes one of the Malay community's supposed lack of cultural-economic networking beyond the national territory. This is what Clive Kessler (1999) has diagnosed as 'diaspora envy', a condition which he considers in relation to Prime Minister Mahathir and the Jewish question. Yet the Malay communal 'lack' might equally have been constructed *vis-à-vis* the triumphalism surrounding overseas Chinese communities' economic success (see Ong, 1997). Kessler shows how diaspora envy has served to stimulate intellectual interest in historical Malay trading networks and contemporary Malay communities overseas. To the extent that the 'new Malay' (*Melayu baru*) is thus (re)imagined transnationally, political space is arguably opened up for Malaysian national identity to be conceptualised more multiculturally. The dilution of 'Malay indigenist' discourses, of course, makes non-Malays less questionable and problematic citizens (Nonini, 1997). On the other hand, such transnational reimaginings undermine the very territorial claims upon which the Malay special political position is founded. Any perceived erosion of this position is hardly likely to be supported by a generation of Malays brought up to believe that their individual and collective economic gains are attributable to economic privileging from the 'Malay-dominated' state. There is therefore an important political geography to Malaysia's multicultural rescripting (see Bunnell, 2002d).

2 Regional imperatives of technological and economic upgrading

Regional exemplars and competitors have compelled technological and economic upgrading as a means of carving a niche in a global(ising) economy. Expansion into high- and information technology sectors has been widely recognised as a vital factor in the economic success of Eastern countries that Malaysia has looked to, such as Japan, South Korea and, more recently, Taiwan (Daly, 1994). The Mahathir era has already borne witness to a significant industrial transition – from dependence on monoculture to an export-oriented manufacturing base. The year 1988 is perhaps significant here marking, as it did, the first time that Malaysia's manufacturing exports surpassed those of primary commodities (Ohmae, 1992). By the time of 'The Way Forward', Malaysia had become the largest exporter of semiconductor chips in the world and, with the launch of MSC, Mahathir affirmed the view that, 'to become a developed country according to our Vision 2020 we cannot continue with conventional manufacturing industries' (Mahathir, 1996a). Although Malaysia had achieved a 'technological deepening' not considered possible in many studies of the New International Division of Labour (NIDL) (Kahn, 1996a: 63), however, microelectronics and other successful manufacturing industries in Malaysia, remained very much dominated by foreign multinational corporations. This applied even in Penang which, prior to the MSC, had been dubbed 'Malaysia's Silicon Valley' (Osman, 1995: 254). However, as early as 1985, The Malaysian Institute of Microelectronic Systems (MIMOS) was established to 'develop a strong indigenous capability in microelectronics and information technology' (*Scientific American*, 1994a: 10).

Efforts to promote the 'indigenisation of technology development' (*ibid.*) were given added impetus in the 1990s. This was no doubt in part motivated by fears of growing competition from other countries entering manufacturing industries: India, China, Vietnam or Indonesia would become relatively more competitive in labour terms, not only because of high wage levels in Malaysia, but also because of its growing labour shortage. Thus, 1992 saw the establishment of the Malaysian Technology Development Corporation to provide venture capital for Malaysian research and development projects. Two years later, Technology Park Malaysia, located on a 314 hectare site south of Kuala Lumpur, provided a local 'home for technology development' (*Scientific American*, 1994b: 12). In line with Vision 2020, these would help foment 'a scientific and progressive society that is innovative and forward looking, one that is not only a consumer of technology but also a contributor to the scientific and technological civilisation' (Mahathir, 1993: 405). Development was thus reoriented to the interrelated production of high-tech spaces and citizens.

The official launch of the Multimedia Super Corridor referred to in Chapter 1 at the opening of the inaugural 'Multimedia Asia' conference and exhibition in October 1996 was therefore a culmination of the increasing prioritisation of 'high-tech'. As alluded to earlier, the launch mapped out a 750 square kilometre site stretching southwards from Kuala Lumpur to the site of the new Kuala Lumpur International Airport. Yet Mahathir made clear that MSC was intended as more than just a 50 by 15 km physical site; it represented part of a broader attempt at 'developing the country using the new tools offered by the Information Age' (Mahathir, 1996a). The MSC, with its physical, soft and information infrastructure, would attract much sought-after mobile investment capital while also helping to create Malaysia's own corporate IT players. If, in the short term, this combination would provide Malaysia with locational advantages in relation to cheap labour competitors, in the long term, it was intended to propel Malaysia(ns) to a higher level of competition altogether.

3 Neo-liberal discourses of globalisation

While a global shift has rationalised the prioritisation of 'high-tech', neo-liberal discourses of globalisation have shaped liberalising means to achieving this in Malaysia. The ideological influence of Kenichi Ohmae, in particular, appears to have been extremely significant. For Ohmae, who was economic advisor to Mahathir in the mid-1980s – and who remained a close confidante of the Prime Minister into the 1990s – economic liberalisation was imagined as the way forward for Malaysia in an increasingly 'borderless world' (Ohmae, 1992). Ohmae was also more specifically involved in shaping Malaysia's decision to pursue technological innovation through his involvement in Malaysia's National Information Technology Council (NITC). MSC, indeed, was said to have been based on Ohmae's plan to make Tokyo into a 'Multimedia Supercity' which formed part of his (unsuccessful) campaign to become Governor of that city (BBC2, 1998).

Mahathir's MSC-launch speech at Multimedia Asia drew heavily upon Ohmae's management vocabulary and conceptualisation of the necessity of liberalisation for a borderless global economy. Although the development of information and multimedia technologies are acknowledged as part of a '*national* strategy' to achieve Vision 2020 goals, the Information Age of which MSC is said to form part is one in which, 'borders are disappearing due to the ease of global communications, capital flows, the movements of goods and people and location of operational headquarters' (Mahathir, 1996a). The borderless world has implications for the nation-state not least because 'where countries once competed, with one nation's trade surplus resulting in another's trade deficit, in the future both countries can benefit ...'. (*ibid.*). There were perhaps also national lessons to be learned from 'mutual benefits' associated with economic liberalisation in the 'borderless economy'.[45] According to Mahathir:

> neighbours prosper more when they help each other than when they are selfish or envious. Sometimes neighbours need new ideas and tools to help them move beyond petty conflicts of the past. These may be frightening at first – because they require fundamental attitudinal changes – but once accepted, people will forget their petty jealousies simply because they are racially or nationally different.
>
> (Mahathir, 1996a)

MSC discourse thus proffered a model of multicultural (trans-)national mutual prosperity.[46] As such MSC discourses are to be understood not only as part of attempts to attract foreign multinationals, building upon economic deregulation in the 1980s (Ong-Giger, 1997), but also in terms of strategies to gain acceptance for a post-NEP stage of post-colonial development.

Optimistic vocabularies of mutual gain glossed over a range of social and political concerns about the desirability of neo-liberal globalisation. Most simply, of course, it was unlikely that Malays brought up in the belief that only state intervention could prevent total Chinese domination of the Malaysian economy would share Ohmae's assertion that 'it's the regulators we have to fear' (Ohmae, 1992: xiv). In addition, as we have already considered, political elites themselves have strong reservations about socio-cultural as well as economic influences from the 'non-East'. Discourses of alternative modernity were precisely a matter of separating the Western 'other' from local/nation/regional modernisation. It is in this light that we can begin to make sense of official descriptions of MSC as a 'test-bed'. Setting aside one 50 by 15 km zone – not only for the application of new information technologies, but also for new forms of cultural-economic governmentality – was thus a spatial manouevre for managing profound reservations about borderless, global transformation.

4 A new urban centrality

In a rather different way, MSC's status as a special 'test-bed' was a sign of a broader (re)centralisation of 'national' development in a period of accelerated globalisation. As alluded to earlier, nation building, particularly from the 1970s, has been oriented towards fostering a more evenly distributed regional development, often understood as a process of overcoming disparities at the level of sub-national states. Homogenising national development practices continued during the Mahathir era, though increasingly by working through planning regions rather than states (Malaysia, 1986). Efforts at dispersion, however, have enjoyed very limited success and industrial development has remained overwhelmingly skewed to the west coast states of Peninsula Malaysia. The dual east–west developmental division evident at independence – between east and west coasts of Peninsula Malaysia and between Peninsula and East Malaysia – has shown considerable resilience during more than four decades of post-independence nation building and economic integration.

It may, in fact, be contended that core–periphery disparities have been extended during the last two decades of the twentieth century. There are a range of possible explanations. One concerns the locational logics of foreign investment. Not only has industrial investment concentrated in and around the principal urban regions best served with internal and international infrastructure – in Penang and, more recently, Johore as well as in the Klang Valley – but a growing tertiary sector has overwhelmingly centred upon the national capital. Geographers have been among those to demonstrate the Kuala Lumpur's domination of producer services (Lee, 1996; Morshidi, 2000). Another explanation for increased (re)centralisation concerns the politics of federal–state government relations. Distant and disputatious states – such as in the east coast of the Peninsula where Islamist electoral opposition has been strong, even winning state governments, as in PAS-controlled Kelantan – have received relatively little federal government infrastructure investment (Bunnell, 2002b).

While infrastructural path dependence and electoral politics are clearly important, however, it is also possible to identify a broader reworking of rationalities of government privileging key urban centres in national development. As considered in the previous chapter, cities are widely understood to have assumed a new centrality in transnational, networked forms of economy. A powerful discourse of producing a 'world class' urban node as a means of plugging nations into an emerging information economy and society legitimises disproportionately high government investment in existing urban centres. In Malaysia, even before the announcement of MSC, monumental infrastructure projects with federal government involvement were focused in and around Kuala Lumpur. The Kuala Lumpur City Centre (KLCC) project and Kuala Lumpur International Airport (KLIA) were retrospectively identified by MSC planners as northern and southern 'nodes' for the new high-tech urban zone. Like the MSC as a whole, KLCC and KLIA were imag(in)ed as *national* projects, essential 'real' and 'symbolic' resources for building the nation-state (cf. Olds, 1995). Yet the spatial concentration of investment that such projects imply points to new socio-spatial division as well as integration in the latest round of national development.

In this chapter, I have reviewed four stages of Malaysian development foregrounding geo-histories of interconnection. In the first stage after Independence, the three generalised political communities of contemporary Malaysia – Malays, Chinese and Indians – which had evolved to service the colonial economy were largely left to their existing socio-economic and spatial positions through *laissez-faire* policies prescribed by nascent supranational organisations and serving largely to sustain foreign economic control. This mode of government came under increasing pressure from Malay economic nationalist demands in the 1960s and ethnic riots at the end of the decade ushered in state intervention to allow constitutionally indigenous *bumiputera* to take their 'proper place' in post-independence national society and economy. If Bumiputeraist policies for a twenty-year NEP 'rehabilitation' of Malays drew upon colonial conceptualisations of evolutionary biology and geographical determinism (Mahathir, 1970), then industrial development from the early 1980s, under the premiership of Dr Mahathir Mohamad, looked East in rationalising further state developmentalism to build the nation and even out socio-spatial wealth disparities across the national territory. The 1990s saw the expiry of the twenty-year policy on affirmative action and the new Vision 2020 rendered visible possibilities for a fourth, post-NEP, stage of national development.

   New rationalities of government during the 1990s were forged in relation to processes of accelerated globalisation, regionalisation and transnationalism. Four key aspects of this for Malaysia's route to 2020 were: possibilities for less Malay-centred state conceptions of national identity emerging from transnational connections and imaginings; an emphasis on technological and economic upgrading as a means of maintaining global competitiveness *vis-à-vis* regional rivals; a perceived necessity for liberalisation in economic and other domains which sits uneasily alongside established ethnic-centred regimes of representation; and a growing emphasis on producing 'world class', connected urban spaces which unsettle homogenising spatial objectives of previous developmental plans. In 1997 'Route 2020' was also the name given to a proposed Dedicated Highway to integrate Malaysia's Multimedia Super Corridor with the existing principal national urban region of Kuala Lumpur-Klang Valley (Multimedia Development Corporation, 1997b: 14). It is to this imagined transect that we turn in Part II of the book, starting in Chapter 5 at the 'northern node' of MSC, Kuala Lumpur City Centre (KLCC).

Part II
On Route 2020

4 Kuala Lumpur City Centre (KLCC)

Global reorientation

The spectacular rise of the Kuala Lumpur skyline in the 1990s was the most visible sign of the city's increasingly global orientation. I have alluded already to the growing role of the city as *the* national centre for advanced producer services connecting it up to regional and global financial markets. Even in the previous decade, architectural observers were reading 'global' influences from the proliferation of internationalist high-rise buildings in the commercial centre (Yeang, 1987).[1] While such signs are clearly significant, however, there is a danger here of underplaying the active role of in situ authorities and everyday inhabitants in fostering *global* urban landscape change. To consider merely how Kuala Lumpur was increasingly subjected to transnational financial flows or marked by the architectural icons of global capital is to posit globalisation as an external force impacting upon an essentially passive city. In the 1990s, Kuala Lumpur, and the larger urban region of which it forms part (the Klang Valley),[2] were characterised by unprecedented attempts by federal authorities to discursively and materially reconstruct urban space and subjectivities in 'global' ways.

A key aspect of this reconstruction in greater Kuala Lumpur, as elsewhere, was large scale infrastructure projects particularly for transport connectivity and the provision of premium commercial office space. The increasingly 'mega'-scale of such projects in East and Southeast Asia in the 1990s meant that they were frequently located on sites beyond existing urban boundaries (Olds, 1995). The 10,000 hectare Kuala Lumpur International Airport, for example, was located at Sepang, some 60 km south of Kuala Lumpur making known a new growth corridor from the national capital. In the 'high-tech' times of mid-1990s Malaysia, it was this area which became the Multimedia Super Corridor (MSC) and so the locus of further globally-oriented infrastructure development. The global reorientation of Kuala Lumpur-Klang Valley was thus bound up with the emergence and official recognition of an extended Kuala Lumpur Metropolitan Area (KLMA) (see Bunnell *et al.*, 2002; and Katiman, 1997). Nonetheless, the single most high-profile project of globalising KLMA was sited near to the existing commercial centre of the city of Kuala Lumpur itself.

In September 1992, Prime Minister Mahathir announced plans for a 'visionary' new city centre project on the 39 hectare site of Kuala Lumpur's former colonial racecourse off Jalan Ampang (Mahathir, 1992) (see Figure 4.1). The plans marked a north-eastward expansion of the city's main commercial district, the so-called Golden Triangle Area (GTA), from Jalan Raja Chulan and Jalan Sultan Ismail. 'Among the largest real estate developments in the world' (*ibid.*), Kuala Lumpur City Centre (KLCC) also came to incorporate a design for the tallest building in the world, the Petronas Twin Towers. In addition to this 452m, 88-storey landmark skyscraper, phase 1 of KLCC included: the Petronas Concert Hall for the newly-created Malaysian Philharmonic Orchestra; Suria KLCC, a six-level shopping mall; a luxury hotel, the Mandarin Oriental; two further office

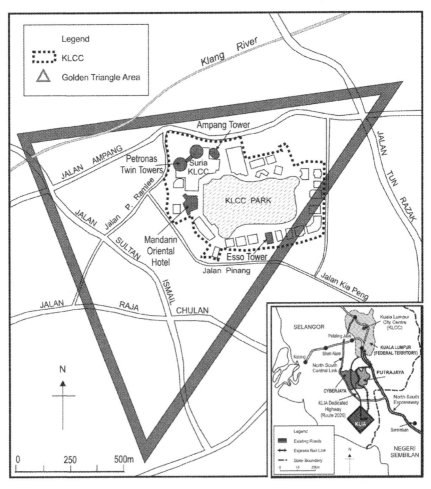

Figure 4.1 Kuala Lumpur City Centre (KLCC).

blocks, Ampang Tower and Menara Esso; a gas-fired district cooling system providing chilled water for the air-conditioning for all KLCC buildings; and a 20.25 hectare public park and gardens which would include a *surau* ('prayer hall') (KLCC Holdings, 1996a).[3] Equipped with 'state-of-the-art amenities and the finest communication facilities', the Kuala Lumpur City Centre Project (KLCC) was subsequently deemed a suitable northern 'node' for the MSC (MSC.comm, 1999a: 20).

KLCC signified city authorities' attempts to transform Kuala Lumpur from national capital to national 'node' in global economic and cultural networks (Bunnell, 2002b; see also Morshidi and Suriati, 1999). Following its formation as a tin mining settlement near the confluence of the Klang and Gombak rivers in the mid-nineteenth century, Kuala Lumpur had grown to become the Selangor state capital in 1880, capital of the Federated Malay States[4] in 1896 and capital of the Federation of Malaya in 1948 (Tsou, 1967). The town emerged from the shadow of Singapore after independence and assumed centre stage in national development following the formation of Malaysia in 1963 (Lim, 1978). In addition to its industrial and commercial dominance noted in the previous chapter, therefore, Kuala Lumpur came to play an important political and cultural role in nation-building (McGee, 1963, 1967). City status was conferred in 1972 and the new city in turn became a separate Federal Territory within the state of Selangor two years later. Authority over the city underwent progressive centralisation: in 1978, the mayor of the city was made responsible to a newly-created Federal Territory Ministry; and nine years later, the authority of the Minister of the Federal Territory was shifted directly to the Prime Minister's Department on account of Dr Mahathir's 'personal interest in the development of Kuala Lumpur' (Phang *et al.*, 1996: 136). It was no coincidence that this period saw the proliferation of 'regionalist' architectural forms, 'indigenous' design references visualising Malay-centred constructions of national identity in state-linked buildings. Yet the very centralisation of administrative power that enabled unprecedented federal government attention on and (politico-cultural as well as financial) investment in the city as a 'national' landscape also had implications for the subsequent transnational or global phase. Mahathir has played a prominent role in the rise of spectacular space in and around Kuala Lumpur, and particularly at KLCC.

In this chapter, KLCC is examined as both representative of and *performative in* Malaysia's global reorientation from the 1990s. In the first section, I consider the Petronas Towers' symbolic function of projecting Malaysia, marking the city and nation on world maps as a modern and 'investible' metropolis. The building was the centrepiece of a capital which was itself a national 'showcase of modernity' (see Evers and Korff, 2000). However, I contend that KLCC also made important internal or domestic performances. In the second section of the chapter, I consider the symbolic intentions of the proponents of the project – to demonstrate that

Malaysia(ns) can compete on a global stage. KLCC was thus not merely diagnostic of an optimistic moment in Malaysian development, but was also intended to provide a new geo-historical vantage point for would-be 'world class' Malays(ians).

Such intended performances, of course, were not passively or uncritically consumed by either international or domestic audiences. In the third part of the chapter, I consider the popular contest and symbolic reworking that attended official (re)presentations of the city during the construction of KLCC. If initial popular critique arose from lack of transparency surrounding the political decision to develop the racecourse site, the rapid rise of the Petronas Towers soon made the building site/sight central to critical views on the nature of urban and national development more broadly. Nonetheless, out of this contest of meanings, it is possible to identify a normative landscape. The global(ising) city landscape, I argue in the fourth and final part of the chapter, has come to play an active role in the production of urban(e) subjectivities. In part, there is a connection here with work on how the scientific ordering of living and working space routinises bodies in the city. However, I focus rather on 'city(scaping) effects': how authoritative conceptions of appropriate conduct shaped ways of seeing and being in the global(ising) city. New Malaysians were induced to realise themselves 'globally' in and through urban space. Globally-appropriate spatial codes were made known through transgression that also rendered those unable to so reorient themselves 'out of place' in a would-be 'world class' city and nation.

Projecting Kuala Lumpur: urban space, global imaginings

KLCC's symbolic role in Malaysia's global reorientation is one of standing for the nation in international imaginative geographies. Clearly, such high-cost real estate developments entail the production of premium urban space for transnational commerce. Yet, across the Asia-Pacific in the 1990s, signature buildings such as the Petronas Towers were also intended to leave a mark in global urban imaginaries – they were built to project a recognisable image of the(ir) city and nation. The aim here was not merely visibility, of course, but rather to be seen in particular, globally-appropriate ways. Gearoid Ó Tuathail has elaborated what he terms a global 'power projectionism': 'the ability of state institutions to initiate and perpetuate hyperreal visions of order, progress and development' (Ó Tuathail, 1997: 313). KLCC is to be understood as part of strategies by national governments, with varying degrees of success, to imagine and configure themselves as 'investible' (Sidaway and Pryke, 2000: 8). However, there was also perhaps a more specifically counter-orientalist re-imaging at work here, opposing: (post-)colonial projections of Kuala Lumpur as a sleepy 'small town capital' (Murphy, 1957: 236) incapable of dynamic development in a region dominated by Singapore (Treggoning, 1966); or else neo-orientalist imaginings of a teeming, unruly metropolis (Bunnell, 2004). Insertion of the KLCC and its landmark Petronas Towers into the skyline of Kuala Lumpur, therefore, was part of a broader symbolic intervention in the imaginative geographies of an imagined Western/global audience.

The investibility and modern urbanity of KLCC were key concerns for proponents of the project. In his short speech at the unveiling of the first phase of KLCC, Prime Minister Mahathir used the phrase *peringkat antarabangsa* on three occasions (Mahathir, 1992) and allusions to the project's 'world class' litter official descriptions during the construction process. As such, Mahathir considered that the project 'will definitely put Kuala Lumpur on the world map' (cited in KLCC Holdings, 1995: 1). While competition for cartographic attention, like that for mobile global capital, undoubtedly takes place at a city or 'nodal' level (see Sassen, 1991; Castells, 1996), however, I contend that the 'world' here was understood primarily as a world of competing *nations*. A brochure on business and investment opportunities at KLCC, for example, described it as 'a project of national significance and a symbol of Malaysia's international status' (KLCC Holdings, 1996a: 23). KLCC, and the Petronas Towers in particular, came to play a symbolic role in 'selling' an urbane Malaysia to the world (Philo and Kearns, 1993).

Architecture and urban design, then, were deployed as signs of national transformation. With its skybridge connecting the twin towers at the 41st and 42nd floors, the Petronas Towers would – in the imagery of the official brochures – define a symbolic 'gateway', not only into the KLCC project (KLCC Holdings, 1996a), but perhaps also externally: to the 'new opportunities' of the developed world. Mahathir took the opportunity at the KLCC launch to describe Malaysia's development as a 'role model'[5] (Mahathir, 1992: np) for other developing nations which, we presume, also wished to bridge the gap between First and Third Worlds. In taking Malaysia 'closer to its goal of becoming a fully developed nation by the year 2020' (Danapal, 1992: 1), KLCC sped the journey out of the developing world.[6] The use of high-rise architecture, in particular, is a well-trodden symbolic path to 'development' (see, for example, Domosh, 1988). The skyscraper has long been imagined as 'a marker of modernity worldwide' (King, 1996: 105).

In Kuala Lumpur, as elsewhere, of course, the socio-spatial impacts of the city's global reorientation were not merely imaginative. A combination of rising land prices for commercial (re)development and a heightened global image-consciousness served to render socially marginal groups increasingly 'out of place' (cf. Cresswell, 1996). Squatter settlements, in particular, were targeted for destruction/development as part of a global 'clean up' of the city (*New Straits Times*, 1997a: 8). Such landscape cleansing was clearly more than merely infrastructural implying, as it did, a new moral geography of who belonged (and who did not belong) in the city (Bunnell, 2002c). Even the apparently empty racecourse site for KLCC was reportedly home to hundreds of horse handlers, workers and their families. These people were relocated to supposedly temporary *rumah panjang* ('long houses') in Rawang, where 39 units made of wood, concrete flooring and asbestos lacked basic facilities such as adequate drainage (Aban, 2001). In addition, squatters were evicted to make way for a new elevated highway that runs into the KLCC site. A pervasive discourse of global(ising) 'improvement' of the city legitimised the eradication of such people and their places throughout the city for other transport infrastructure (Barter, 2002) and modern housing (Yeoh S. G., 2001) as well as for high-rise commercial development.

Yet, as alluded to earlier, Kuala Lumpur has featured high-rise buildings since long before the globalising 1990s.[7] Two characteristics distinguished the Petronas Towers from much of the previous high-rise architecture comprising the Kuala Lumpur skyline. The first concerned the putative attempt to create a recognisably 'Malaysian' skyscraper by incorporating 'local' design features. The towers' architect, Cesar Pelli had previously designed towers in New York, London, Tokyo, Mexico City and Buenos Aires, and so claimed considerable experience of 'tailoring' designs to suit cities and countries with wildly differing characteristics (Moore, 1994). The floor-plan of the Petronas Towers was said to be 'based on Islamic geometric traditions' (Cesar Pelli and Associates, 1997: 29) and the building's design features were intended to 'convey a specific sense of their tropical locale' (*ibid*. 29). The project brief elaborated how the multiple facets created by the form of the towers 'reflect the sunlight forming the combination of deep shadow and dazzling brightness that might be found in a tropical forest' (Cesar Pelli and Associates, 1994). One architectural commentator's suggestion that this symbolism had 'all the profundity and insight of an in-flight magazine' (Moore, 1995: A5) is perhaps instructive. An article in Malaysian Airlines' 'Wings of Gold' noted how 'Big Ben, Eiffel Tower, the Pyramids, even the Little Mermaid conjure immediate images of a particular city' (Kee, 1993: 54). Figure 4.2 picks out the Petronas Towers in the world of monuments traversed by Malaysia's national airline.[8]

The issue of profundity here misses the point. What Mahathir considered was being constructed was a 'cultural landmark' (Mahathir, 1992: np) which would allow the city and nation to 'travel'. It is specifically the *image* of the Petronas Towers which was used in 'advertising the country's arrival as a modern industrial nation' (*Progresssive Architecture*, 1995: 86). As Anthony D. King has pointed out,

> in what is now a totally institutionalised mimetic televisual convention, it is the White House, the Houses of Parliament, the Duma or Eiffel Tower which – subliminally elided into the capital city – is used to mediate the meaning of the Nation to the gazes of the World.
> (King, 1996: 101–2)

When it is not the individual building which fixes the global gaze on the rectangular screen, it is a wider view of the city skyline (as in Figure 4.2).

Figure 4.2 We fly the world.

Source: Reproduced with permission from Malaysia Airlines.

The Petronas Towers thus flagged Kuala Lumpur and Malaysia as part of a 'global skyline' (King, 1996: 100): 'this unique pair with the distinctive outlines, is a national landmark in a global setting' (KLCC Holdings, 1996a: 7). There is, of course, something of a paradox in a situation where the material and imaginative work of constructing the 'uniqueness' of place is expressed through a universally-understood language of post-modern architecture, couched in a universal English language marketing vocabulary[9] and performed by a global network of property development 'experts' (Olds, 1995, 2001).[10] Nonetheless, as for other cities, a landmark image travels in extra-televisual ways: on postcards, t-shirts and other memorabilia as well as in-flight magazines and advertisements (see Monnet, 2001).

Earlier attempts to create supposedly recognisably 'Malaysian' skyscrapers should also be noted here. These emerged from the 1980s partly as response to government demands that design professionals formulate a national architectural aesthetic (Tay, 1989). Prime Minister Mahathir himself reportedly saw no reason why 'a skyscraper should not have a roof which reflects our national identity' (cited in Kultermann, 1987: 68). Reflecting the coincidence of these demands with the height of the National Cultural Policy, state-sponsored architectural nationalism came to mean the incorporation of specifically 'Malay' symbols into modern building forms. As alluded to above, therefore, the proliferation of high-rise landscape artefacts with 'indigenous' features were bound up with the construction of a Malay-centred national identity. Two such high-rise buildings in Kuala Lumpur are Menara Maybank (the roof of which is shaped in the form of a Malay *keris* ('dagger') and the 'LUTH' Tower in which five columns are said to represent the five pillars of Islam.[11] Subsequent regionalist architecture in Malaysia has, however, become more sophisticated. Ken Yeang, for example, has considered the functional basis of a 'bio-climatic' skyscraper for a tropical environment (Yeang, 1994). For Yeang and others, many earlier regionalist experiments amount to little more than conventional 'temperate' designs with supposedly local motifs 'tacked on' (Yeang, 1987).

In this light, the award of the monumental KLCC project to Cesar Pelli was a triumph of global architectural image over tropical regional bio-climatic function.

The second way in which the Petronas Towers was distinguished from previous high-rise structures in Kuala Lumpur – including other national or regionalist ones – was simply a matter of how high. The Petronas Towers rose to become 'head and shoulders' taller than the mostly 'placeless' high-rise buildings of Kuala Lumpur's Golden Triangle Area. When the building was 'topped out' in 1996 it also claimed the world building height record from Chicago's Sear's Tower, if only by means of contentious functionless spires: 'Chicagoans' reportedly appealed to the Council on Tall Buildings and Urban Habitat for a 'hats-off' definition (Kaur, 1996). By this time, Mahathir actually seemed to play down the record-breaking height of the Petronas Towers. This apparent modesty may be explained by awareness of plans in other cities, especially in Asia, to build even higher (*Business Week*, 1994).[12] The ephemerality of 'symbolic capital' applies as much to the 'global skyline' as to that of any individual city (Dovey, 1992: 186).[13] There is nothing especially thrilling about being the fourth, third or even the second tallest building in any 'world'.

The record stood, conferring the towers, Kuala Lumpur and Malaysia with considerable international visibility and upward mobility. The international press in the 1990s was awash with building height charts which replicated evolutionary biological diagrams with a new architectural geography of tallest and fittest (see Figure 4.3). Certainly, the symbolic statement of constructing the world's tallest building on a colonial racecourse – combined with regionalist assertions of 'Asia' as the new leading edge of modernity – was not lost on the former imperial centre. Reports on the Petronas Towers in the British press discussed the building in terms of other past and projected record-breaking attempts. There was, no doubt, good journalistic reasoning behind this – putting the building 'into context', historically and/or spatially. Yet it is significant that this contextualisation was frequently concerned less with city- or national-level transformation and competition than with the emergence of a dynamic Asian region.

Contesting the fashionable notion in the 1990s of a new Asia(-Pacific), which was threatening the West's position at the leading edge of modernity, Anthony D. King argued that 'nations in Asia may well be entering a global competition which most of the other contenders have already abandoned' (King, 1996: 111). He drew, in particular, on an article in *Progressive Architecture* to show how the rash of record-breaking buildings in Asia was following architectural models and principles now largely discredited in 'the West'. The very fact that so many of the design contracts for these megaprojects, including the Petronas Towers, were awarded to American companies served to undermine claims of the eclipse of Western scientific or economic superiority.[14] But, more than this, American companies were said to be practising in Asia (what in a 'Western' context would be considered) 'socially irresponsible architecture' (*Progressive Architecture*, 1995). The implication was that urban modernity of/in the West had moved on and that Asians were merely importing an 'obsolescent modernity' (King, 1996).[15] While there was a danger here of mapping Asia into a singular geo-historical modernity centred in North America, King was surely correct in suggesting that 'the public display of sheer size serves less to demonstrate its superiority than to confirm the consciousness of its inferiority' (*ibid.* 104).

It is also possible to cast doubt on the success of these record-breaking attempts in cartographically (land)marking specific cities or nations. Writing about the high-rise boom in Asia, one architectural commentator made the point that 'the world's tallest towers are planned for cities few in the West could place, let alone pronounce' (Sudjic, 1996: 1). To this may be added, spell: another article located the world's tallest structure in 'Kuala Lumpa' (Jury, 1996: 5). To what extent did the Petronas Towers serve to put Kuala Lumpur or Malaysia on world maps? Neo-orientalist reporting subsumed Kuala Lumpur, Malaysia (and other unspellable, unpronounceable places of uncertain sub-regional location) within a rising Asia. Yet this was an Asia whose defeat of America at its own architectural game was read as a sign of a broader challenge to 'the West' (*Progressive Architecture*, 1995: 44).[16]

In the mid-1990s, therefore, KLCC had symbolically projected Kuala Lumpur and Malaysia as part of a miraculous Asia, if not entirely asserted its symbolic centrality to this fast-modernising region. On the one hand, there was clearly commercial benefit to being cartographically imagined in terms of an economically-dynamic region. The project was said to be a 'visible and viable commercial enclave for a great city in the fastest-growing region in the world' (KLCC Holdings, 1995: 4). KLCC marketing thus drew upon, and contributed to, the discursive construction of a fast-rising economic region. Yet it also sought to mark the project's positioning as a 'hub' for that region. Thus, on the other hand, KLCC was promoted as '*the* business location in South-east Asia' (*ibid.*, 1995: 1). The notion of KLCC as 'gateway' was intended to evoke not so much regional borders or boundaries, but suggested an invitation to its centre – to a project 'strategically located in the heart of South-east Asia' (*ibid.*, 1995: 6). It is the domestic effects of such globally-oriented urban imaginings which forms the focus of the next section.

Symbolic intentions: new perspectives for new Malay(sian)s

The reorientation of Kuala Lumpur in the 1990s was not reducible to attempts to connect up to a global modernity centred elsewhere. It is also important to consider a context of increasingly self-confident political discourses positing 'autonomous' definitions of modernity (Ong, 1996). Proponents of the KLCC project clearly saw their developmental contributions in terms of regional- (Asian, Southeast Asian, East Asian, Asia-Pacific, etc.) as well as national-scale modern transformation. The aim here, however, is not to identify a particular geo-historical modernity (or modernities) of which the KLCC project is somehow representative or diagnostic. Rather, I am concerned with how the site/sight itself was actively bound up in the (re)formation of geographical subjectivities in 1990s Malaysia – new ways of seeing, and being in, the world.

As we have considered in previous chapters, the economic success of nation-states such as Malaysia emboldened political elites and cultural authorities to articulate their own visions of modernity. The location of the world's tallest building in Kuala Lumpur added credence to such imaginings. Not only had Malaysia and other economies in a rising East achieved a Rostovian 'take off', but the evolutionary charts of socio-economic 'progress' which had staged the West as the end-point of being modern now continued their ascent to a new apex of A(/Malay)sian development. Thus the building height charts so popular in Malaysia and the international media in the 1990s appeared to corroborate the civilisational arguments of Mahathir and other proponents of 'Asian values'. Figure 4.3, for example, could be read in terms of a new civilisational stage geographically centred in Malaysia: from the 'ancient world' of Egyptian civilisation (the pyramid); to the 'old world' of Christendom (the cathedral); to the beginning of (what some would consider to be) the 'modern world' following the Industrial Revolution in Europe (the Eiffel Tower); to the 'new world' of US corporate and ideological dominance (skyscrapers); and, ultimately, to the Petronas Towers taking its place in the sun.

While the high-profile charting of KLCC entangled the site/sight in a myriad of contesting and often contradictory imaginings, the national

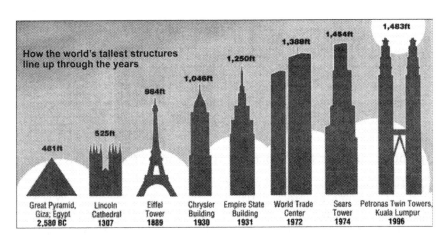

Figure 4.3 High and mighty: how the world's tallest structures line up through the years.

Source: © *The Guardian*, March 1996. Reproduced with permission.

cultural and political significance of the project should not be underestimated. Earlier 'national' high-rise buildings, as we have seen, were a product of a 1980s architectural regionalism that coincided with the height of the Malay-centred National Cultural Policy. Given the increasingly multi-cultural state rescripting of national identity in the 1990s, the explicitly 'Islamic' symbolism of the Petronas Towers was perhaps surprising. One British architectural commentator suggested that this putatively national symbolism 'will mean little to the large number of non-Islamic Chinese' (Moore, cited in Turner, 1997: 36). The role of the national oil company, Petronas, is clearly important here. It was, in part, the extraction of newly-discovered petroleum reserves and rising oil prices from 1973 which had financed public sector growth for NEP affirmative action on behalf of the Malay community (Jomo, 1995: 4). Petronas' previous headquarters, the Dayabumi Complex, also featured an ostensibly 'Islamic' façade. For Abidin Kusno, the Petronas Towers were intended to symbolise the *Melayu baru* ('new Malay') entering the world of commerce (cited in King, 1996: 113).

While the Islamic façades of 'national' towers in the 1980s were unambiguously Malay-centred, Petronas' *Melayu baru* could be reconciled with an increasingly multicultural state rescripting of national identity and the 'global' transformation of which this formed part. The new Malay was globally-oriented, forward looking and looking forward to a Malaysia beyond constructive protection. As such, apparently ethnically-exclusive architectural references could also envision a national future of multicultural tolerance and complementarity. The rise of the *Melayu baru* in the corporate landscape of Kuala Lumpur, in other words, did not necessarily signal the economic and/or cultural displacement of national 'others', but opened a symbolic space for the enfolding of Malayness into a broader *Bangsa Malaysia*. Somewhat paradoxically, therefore, the necessity of and will-ingness to change was imagined through the architecturally familiar. Cesar Pelli, who also designed Canary Wharf in London, compared (what he saw as) receptiveness to the Petronas Towers in Kuala Lumpur with negative reactions to 'the first true skyscraper in England': Canary Wharf sits uneasily at the edge of a city which is very ambivalent about skyscrapers, whereas, 'the Petronas Towers are for a city that is embracing them wholeheartedly'. Malaysia, he added, 'sees itself as moving forward to the future whereas some think that Britain had its best days in the nineteenth century' (cited in Moore, 1994: 21).

Comparison with the UK is instructive in other ways since the wholehearted acquisition of the latest technology in Malaysia has been compelled politically as an imperative of 'resistance' to (neo-)colonisation. For Malaysia, as Wendy Mee has suggested, colonialism was 'an early lesson in vulnerability in the face of superior technology' (Mee, 2002: 61). A belief in the importance of science and technology was a prominent theme in proto-nationalist writings from at least the mid-nineteenth century and remained integral to post-colonial Malaysia's global shift in the 1990s. As Mahathir confirmed in 1997, 'the acquisition of knowledge and sophisticated technology which is no less than the developed countries' is necessary to 'prevent from being colonised in a new way' (Mahathir, 1997a). One text on Mahathir's 'leadership' in science and technology in Malaysia, suggested that the transition to becoming a 'truly sovereign people' demanded 'minds and institutions that can inquire, assimilate, innovate and reformulate knowledge' (Abdul Aziz and Pillai, 1996: 3). The skyscraper, of course, has been understood as a celebration of modern building technology (Huxtable, 1984). From the outset, KLCC was lauded in terms of its technological sophistication. In an ironic coincidence, the early stages of construction coincided with a trough in relations between Malaysia and the former colonial centre. Following a British newspaper report alleging corrupt practices in another large infrastructure project, the Pergau Dam in northern Peninsula Malaysia, Mahathir defended national credibility against what he cast as neo-colonial meddling in Malaysian affairs. The resultant ban on British companies from tendering for public sector contracts, especially lucrative infrastructure projects, signalled state determination to (be seen to) realise their own technological futures (Dynes, 1994).

Promoted as a benchmark for national technological achievement, the Petronas Towers starred in advertisements for a range of modern Malaysian commercial and industrial innovations, from trucks and telephones to refrigerators. Figure 4.4 shows MEC's '3-door frost-free refrigerator' in an imagined skyline comprised of Malaysia's other 'greatest achievements'. In addition to the Petronas Towers, these include (on the left) the tallest flagpole in the world, which is in Merdeka Square, and (on the right) Menara Kuala Lumpur, the world's 4th tallest telecommunications tower. As can be seen, the advertisement also includes a semi-diagrammatic representation of MEC's refrigerator range which increases in height (and volume) from left to right, thus evoking charts of building height increases discussed above (see Figure 4.3) and of evolutionary progression more broadly. Unlike the Petronas Towers, the 3-storey 500 litre model may not have been the tallest in the world, but it was 'the largest ever to be made in Malaysia'. What is more, designed to 'international quality', MEC's fridges – like *Melayu baru* – were equipped to 'stand up' to the competition of the 'world market'.

Notions of corporate and industrial competitiveness formed part of a broader post-colonial sense that *Malaysia boleh* ('Malaysia can do'). One article in the international press suggested that 'the Towers reflect the current mood of expansion where high rise seems to equate with high ambition. They tune in to the spirit of the age' (Moore, 1995: 8). However, I contend that the construction of the building in the 1990s did not simply 'tune into' or 'reflect' a mood of optimism. Proponents of the Petronas Towers intended them as Malaysians' 'towers of strength' (Bowie, 1997: 8),[17] proving to the population that the sky is the limit for what they 'can do'. The multicultural Malaysian Everest Team's conquest of the world's highest

MALAYSIA'S GREATEST ACHIEVEMENTS

HRF-C530FA

Flip-Top Ice Box   Multi-Air Flow

3A - 100% CFC-FREE
MEC
WE CARE FOR THE OZONE

Deodoriser   Chilled Room   High Gloss Door

Presenting a Malaysian breakthrough in refrigeration technology. The MEC 3-door frost-free refrigerator, model HRF-C530FA – with a host of attractive new features.

With big, roomy 500 litres of storage space for all the needs of your family. It's also the largest ever to be made in Malaysia.

Many user-friendly features have been specially added to give you extra convenience. Among them: Flip-top Ice Box for easy removal of ice cubes, Multi-Air Flow to ensure the freshness of food in every shelf while the Deodoriser reduces mingling of smells.

Then, there's the unique long-lasting High Gloss Door that allows you to clean with greater ease. No detail is too small for us. Like the Roller Wheels that help you move the refrigerator to a new position in your house.

Behold this technological beauty – MEC's international quality refrigerator. Made by Malaysians for the world market.

There is also a wide range of refrigerators to choose from. And for every model, the same MEC's quality excellence is there to serve you, year after year.

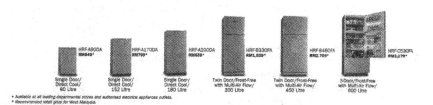

HRF-A90DA RM648*   HRF-A170DA RM799*   HRF-A200DA RM839*   HRF-B330FA RM1,829*   HRF-B480FA RM2,709*   HRF-C530FA RM3,179*

Single Door/   Single Door/   Single Door/   Twin Door/Frost-Free   Twin Door/Frost-Free   3-Door/Frost-Free
Direct Cool/   Direct Cool/   Direct Cool/   with Multi-Air Flow/   with Multi-Air Flow/   with Multi-Air Flow/
60 Litre   152 Litre   180 Litre   300 Litre   450 Litre   500 Litre

* Available at all leading departmental stores and authorised electrical appliances outlets.
* Recommended retail price for West Malaysia.

Figure 4.4 Malaysia's greatest achievements.

Source: Reproduced with permission from MEC.

peak in May 1997 was described in a rash of media coverage as 'an achievement that shows the world Malaysia Boleh!' (Telekom Malaysia and Everest '97 Team, 1997: 7). In a similar way, KLCC contributed to the fostering of a sense of 'world' recognition of Malaysia's national achievement.[18] Indeed, as we have seen, the record-breaking height of the Petronas Towers in particular was more likely to attract the gaze of the world since, unlike Everest, the building represented a peak that no other nation had previously reached. 'Well, KL has won' proclaimed one newspaper headline on international recognition of the city's height-breaking 'strides forward' (*New Sunday Times*, 1997a: 10). The way others see – or were said to see – Malaysia added 'global' legitimacy to new ways in which Malaysians were intended to see themselves.

If the Petronas Towers, like the summit of Mount Everest, was a vantage point for Malaysia's ascent to *peringkat antarabangsa* then we begin to see the constitutive work of KLCC in the making of new Malays(ians). The project legitimised and compelled socio-spatial change, realising new ways of seeing and being in the world. Yet I have alluded already to how global landscapes also marginalised people and places considered less than 'world class'. It is no surprise, therefore, that the symbolic intentions of KLCC's proponents did not translate unproblematically into new citizen subjectivities.

Symbolic discontent and popular reworking

The intended symbolic meanings of KLCC were subjected to reworking and contest as everyone came to have a view on the Petronas Towers. The building, whether through direct urban encounters or reproduced in a variety of media, was enfolded into the practice of individual and collective lives. Geographical subjectivities were thus forged in and through the building which, in turn, accumulated new layers and levels of meaning. Clearly this proliferation, reworking and accumulation of meanings implied a complexity of subjectification which extended beyond a mere internalisation of official perspectives or values. Yet it is also important to avoid counter-posing diverse and messy lived realities with a supposedly coherent or unified authoritative intentionality. Official representations of the building were themselves diverse and dynamic. Nonetheless, the high-profile of the Petronas Towers nationally and increasingly – with each new story/ storey – within the city, made it the target for symbolic discontent. In the context of an entrenched ruling coalition with little tolerance for criticism and a generalised lack of public participation in the planning process, such discontent may be understood as a significant 'popular politics'.

The Selangor Turf Club was a site of public contest even prior to the official announcement of KLCC. Plans to relocate the Jalan Ampang racecourse had, in fact, been announced as far back as August 1982, but this was said to be in order to turn the site into a 'park for the public' (Padman and Lim, 1989: 8). Following the imposition of Section 8 of the Land Acquisition Act in October 1983, the Selangor Turf Club (STC) was given three years to find an alternative location before its 39 hectare site was converted into the park. City Hall eventually issued an eviction notice in May 1988, but STC was granted another three years to develop a new race track at Sungei Besi, Selangor. STC, meanwhile, proposed that it would cede its 27 hectares of leasehold land to the government without compensation in return for permission to develop the remaining 12 hectares (which was freehold land) for commercial purposes (Padman and Lim, 1989). However, following an extraordinary general meeting on 18 August 1989, STC announced the sale of all 39 ha to Seri Kuda Sdn Bhd raising fears that, rather than being 'opened' to the public, the former colonial racecourse would be developed into real estate as part of the city's exclusive Golden Triangle Area (GTA) (*Sunday Star*, 1989: 6). Seri Kuda's application to rezone the site was publicly opposed by environmental and heritage groups.

Public opposition was channelled into proactive idea(l)s for alternative uses of the land. An architect, Ruslan Khalid, wrote a letter to the *New Straits Times* (NST) proposing a public donation fund, 'so that we can create a park for the people from money raised by the people' (Ruslan, 1989: 15). Some months later, an article in the NST proposed a possible plan for a 'People's Park' under the heading 'this land is your land' (Edwin, 1990: 9). The accompanying text stressed that it had only been because the club was 'exclusive' that the public had failed to recognise the potential of the land to become 'an oasis in an asphalt desert'. This oasis would provide not only shelter from the 'madding crowd' (for which a 'Dome of Serenity' was provided), but also the 'experience of nature' in a Federal Territory described as 'growing into a concrete jungle'. In response to the alienating 'harsh effects of towering concrete', a windmill was incorporated in the plan to symbolise 'the harmony of life and the generation of pollution-free energy'. An Artist's Corner and Graffiti Wall would allow creative self-expression. The following year, residents of Kuala Lumpur learned that 'their' land would be developed as a RM 7 billion city centre (*New Straits Times*, 1991: 4).

The former racecourse, then, had become a stage for contest over definitions of appropriate urban development. In part, resistance to the development of the site relates to broader critiques of global capitalism and, in particular, the potentially deleterious environmental and social effects of the determination of land-use by market value alone.[19] For some, KLCC signified the spatial reorganisation of the city, and the GTA in particular, by market principles and the broader privatisation of space in Malaysia (Sobri, 1985).[20] In addition, however, the decision to proceed with KLCC prompted critique of specifically-Malaysian developmental practices and attitudes. One environmental NGO (Non-Governmental Organisation) informant outlined examples of a 'lack of accountability' in the KLCC project: over-riding the allocated land-use for the racecourse on the Kuala Lumpur

Structure Plan against public wishes; the fact that no Environmental Impact Assessment (EIA) was carried out; and failure to make public the results of a Department of Environment traffic survey.[21]

Critique of the decision to redesignate the racecourse, against the wishes of 'the people' may be understood in terms of a broader discourse of 'accountability' in the 1990s (see Harper, 1996). Seventy per cent of Seri Kuda, which purchased the real estate 'plot of gold' (Ho, 1991: 15), was owned by Pacific States Investments Ltd., reportedly controlled by T. Ananda Krishnan (Yong, 1991), a former director of both Bank Negara and Petronas and long-time associate of Mahathir (Gomez and Jomo, 1997). The KLCC project thus confirmed popular belief that all major decision-making in a hierarchical process emanated from the Prime Minister himself and that even private sector success was dependent on close relationships with the upper echelons of UMNO. Criticism of KLCC within UMNO was most likely because Krishnan is a Sri Lankan Tamil, not 'Malay' (Gomez and Jomo, 1997), but the involvement of Petronas, in particular, also evoked earlier suggestions of high-level misuse of public funds. In the case of the Dayabumi Complex – another 'Mahathir project' in the 1980s – Petronas had been used to rescue the owners, the Urban Development Administration, by paying almost twice the market value for the building (Bowie, 1991).

The final plan for KLCC included a twenty hectare park (KLCC Holdings, 1996b). Designed by the 'award-winning' Brazilian landscape artist, (the late) Roberto Burle Marx, the park, at one level, contributed to KLCC's international real estate appeal. Marketing literature envisaged the park as 'balm for the stresses of work' (KLCC Holdings, 1996a: 9). However, the park might equally be read as a response to public opposition to the commercial development of the racecourse site. T. Ananda Krishnan described it as 'a gift to the public' (Ho, 1991: 15). Official marketing literature appropriated much of the vocabulary of previous contest: KLCC park was to become a 'green lung freely accessible to the public' and 'a park for the people' (KLCC Holdings, 1996b). When rumours of redevelopment first reached the press, the minister in charge of the Federal Territory and its development had reportedly warned: 'city folk have been waiting for this park since the idea was mooted in the early 80s. I think any plan contrary to this will disappoint the public' (Dr Sulaiman Mohamed, cited in *Sunday Star*, 1989: 6). While the public was not ultimately disappointed in terms of the provision of a park, lack of transparency in the process through which KLCC had come into being had fomented popular rumour and gossip by the time construction began.

Two rumours, in particular, are worthy of note. The first concerned who was *behind* the project, its financing and the role of Petronas in particular. The oil company's rescue of the Dayabumi Complex had occurred as recently as 1989 and yet this building was to be vacated so that Petronas could become the 'anchor tenant' for the twin towers. It was initially thought that Petronas and all of its subsidiaries, housed together in one building for the first time, would occupy one-and-a-half towers, or three-quarters of the total floor space. This was later seen to be an over-estimate, leaving the whole of Tower Two as lettable space.[22] For all the undoubted attractions of the building in terms of both facilities and prestige, which other companies would be willing to 'pay the price' for building so high?[23] Others were simply concerned with finding out more about 'the enigmatic AK' (T. Ananda Krishnan) (*Malaysian Business*, 1991: 15). A second rumour related to what was *beneath* the project and the Petronas Towers in particular. An article in *Building Property Review* eventually confirmed that the towers had been moved 50m from their original location on account of limestone caverns close to the surface. At the final location, a grouting programme was carried out to fill cavities in the limestone with cement and to improve the slump areas (*Building Property Review*, 1995). Nonetheless, one critical commentator still asked 'why anyone would want to build anything so massive on an area notorious for sink-holes' (*Aliran Monthly*, 1996: 19). Attempts to get a clear view of the project during the early stages of construction were also more literally hampered by the hoardings around the site which reduced its development to the mumble of machines and voices of foreign workers.

As the twin towers grew visible they became the focus of further symbolic reworking. The 'race to the top' (*Business Times*, 1996: 2) between rival Japanese and South Korean-led teams working on Towers One and Two respectively suggested that this was a project which, like its principal proponent, was 'in a hurry' (Khoo, 1995). Construction had proceeded in the face of public objections and apparently without sufficient geological research. For those already convinced of Mahathir's super-humanity, it was perhaps only the force of the Prime Minister's personality which was keeping the building upright. Popular organic comparisons – to an upturned *jagung* ('corn cob'), for example – may be said to have inferred a conception of the building's emphemerality, its perishability. All would not have been lost in the event of collapse, however, as Malaysia might still have been home to the longest building in the world![24] Dr Mahathir's building was repeatedly diagnosed as symptomatic of a 'national obsession' with superlatives (Mustafa and Subramaniam, 1995: 5). The tallest flagpole in the world graced the Kuala Lumpur cityscape before the 'sky-pricking' fantasies of the Petronas Towers (Jencks, 1980).[25] Few were surprised, then, when these really were joined by plans for the world's longest building. Stretching twelve kilometres along the Klang River, Kuala Lumpur Linear City (KLLC) would be far longer than a horizontal Petronas Tower (or two) (see Marshall, 1998). Even the re-siting of the Petronas Towers and remedial strategies for the cavernous limestone foundations were glossed in terms of 'the single biggest Concrete Pour for a building foundation in the world to date' (*Building Property Review*, 1995: 58). The emphasis on quantity, speed and size did little to inspire confidence in the quality of the construction and perhaps helped people to perceive that Tower Two was leaning (Murugasu, 1999).[26]

The sheer size of the building, however, ensured its centrality to many views on the city and its global redevelopment. As for Roland Barthes' description of the Eiffel Tower in Paris, so for the Petronas Towers in 'KL': there was no 'fear nor fantasy' which failed, sooner or later, 'to acknowledge its form and to be nourished by it' (Barthes, 1979: 4). Unfortunately, in the construction stage, at least, KLCC did little to ameliorate everyday lives of most city residents. On the contrary, in addition to reducing Kuala Lumpur's already sparse 'green space', the development of KLCC only served to augment air pollution and traffic congestion in the capital (*New Straits Times*, 1996a: 22). It was no coincidence that Jalan Ampang became a renowned traffic-jam black spot in an already-congested city. These construction-stage problems fed fears that completion – with underground parking for 5,000 cars and with up to 5,000 workers employed in each tower – would only mean more congestion.[27] The likelihood of looking favourably upon the towers was perhaps also undermined by their own militaristic stance. 'The world's tallest buildings', wrote Malaysian poet Cecil Rajendra in his 'War Zone', 'look like giant shell casings, embedded in the heart of our city' (Rajendra, 1999: 9). With its armour-like steel cladding,[28] 'fortress KLCC' ushered in the 'militarisation' of urban space (see Davis, 1992) with 'boardroom battles' and city life as a 'battle-zone' (Rajendra, 1999: 9).

The front cover of a book by Malaysia's most famous cartoonist, Lat, encapsulated further popular problematisations of the development of Kuala Lumpur (Lat, 1996). A new arrival to the city sits in his Proton at the end of an uncompleted road. Smoke billows from the on-going construction work (which is also suggested by a crane on the skyline) reducing the traffic below to a standstill. Even in *Lat Gets Lost*, however, there is perhaps a sense of problems relating not only to incompletion, but to the state project of development more generally. The driver on the cover is faced with the choice of turning around – leaving the 'development' of the city behind – or else plunging over the edge. His map provides few clues as to alternative routes; the cartographer's view is as out of touch with the street level(s) as are the Petronas Towers and Menara Kuala Lumpur shown on the skyline. The Master Planner's panoramic 'totalisations' (see de Certeau, 1985: 124) could not discern the fine detail of everyday lives which consisted of getting lost (the Proton driver) or else going nowhere (the stationary traffic below) (see also Bunnell, 1999).

The haze which afflicted Kuala Lumpur and much of the rest of Southeast Asia during fieldwork in 1997 added another dimension to metaphors of vision/sight in popular problematisations of city development. Fears for the health of citizens in the smog which worsened from August (1997) led the state Meteorological Services Department to publish daily recordings of an Air Pollution Index (API) in different parts of the country. However, some doubted the accuracy and even honesty of figures associated with image-conscious authorities (Pillai, 1997). It was the Petronas Towers which came to be (not) seen as huge 'twin dipsticks' for air pollution levels in the city (Sia, 1997a: 4). A contributor to an Internet forum described an acquaintance's 'personal index' as follows: 'if the Petronas Twin Towers can be seen in outline in a haze, it [the API] is over 200. If it cannot be seen it must be over 250. If the neighbouring buildings are also encased in smoke, more than 300'. Not only this, but the Petronas Towers method 'is more effective than the figures the government gives out' (Pillai, 1997). Severe air pollution, whether attributed to industrialisation in Peninsula Malaysia itself or else understood in terms of forest fires in Kalimantan (Indonesian Borneo), fomented critical views of Malaysian development more broadly. Forty three Malaysian companies were reportedly involved in the logging in Kalimantan, some of which were state-owned (Vidal, 1997; Kua, 2002).

Further popular reworking concerned less the visibility of the building than whether it could see them. Rumour had it that Prime Minister Mahathir himself would occupy the top floor of one (or both) of the towers. Why was there no lettable space above the eighty-fourth floor in the 88-storey towers?[29] The fact that the top floor of each tower became home to devices for cleaning the façade of the building did little to diminish a popular perception of the state keeping its eye on the city population. Even in those moments when the Petronas Towers disappear from view in the national capital, there is a sense that the building – a building very much bound up with state power – is watching you.[30] Official literature on KLCC promoting its 'intelligent buildings' perhaps only served to confirm suspicions of surveillance (KLCC Holdings, 1995: 4).[31]

By the time they were 'topped out' in March 1996, everyone had a view on the Petronas Twin Towers. Clearly this was not limited to those living and working in Kuala Lumpur but to more distant others, nationally and even beyond. Just as intended meanings were contested and reworked through experiences in the city, KLCC's official 'travel' through glossy investment brochures and leasing literature was mirrored by symbolic discontent and popular reworking 'at a distance'. This ranged from purely playful re-presentations in established media to more self-consciously critical contest in burgeoning electronic spaces. Nonetheless, through whatever media, KLCC had been enrolled into lives in the city and elsewhere – many views had been changed by/through the project. It is to the cityscaping of new ways of seeing and being to which I now turn.

City(scape) effects: a new urbanity

If KLCC produced new ways of seeing and being that were not reducible to the symbolic intentions of the project's proponents, then we need to think more carefully about the subjectifying effects of urban sights/sites. How was the global reorientation of Kuala Lumpur through projects such as KLCC bound up with new ways of city living and working in the 1990s? At one level, the constitutive work of space may be understood in terms of the role of urban planning and architectural design as technologies of bodily disciplining and routinisation. Jean Gottman famously outlined the efficiency of work in the 'vertical system' (Gottman, 1966) and others have considered applications of scientific management in high-rise residential developments such as condominiums (e.g. Dennis, 1994). In this section, however, I consider not so much how subject-positions are shaped by architectural design as how landscape subjectification works through embodied normative judgements in urban space.

To the extent that certain modes of conduct are considered desirable or appropriate in the global(ising) city, space itself came to 'perform something' (see Razack, 2002: 9) in the shaping of urban(e) subjectivities. Two points of clarification are necessary here before I proceed to elaborate city(scaping) effects in Kuala Lumpur. One concerns delimitation or definition of the urban(e) social 'code'. What might be understood as the normative landscape is a contested and dynamic product of a diversity of authorities, from management gurus to design professionals to social commentators and more conventional 'political' figures and institutions (see Bunnell, 2002c and Cresswell, 2000). A second point of clarification concerns the limits or limitations of authoritative code(s). Far from translating unproblematically into a new urbanity, appropriate conduct was often made known precisely in opposition to those unable or unwilling to realise themselves in 'world class' ways.

Melayu baru (the 'new Malay') mentioned above provides a useful way into thinking about how appropriate new forms of conduct are realised in and through city space. As we saw in the previous chapter, from the 1970s in particular, rural–urban migration was promoted by the state and other authorities as a means of Malay modernisation. A simple environmental determinist faith in the accelerated developmental effects of the city, however, was clearly a flawed strategy for Malay cultural-economic transformation. Rather than leaving the *kampung* behind, migrants imported it into the city or, more frequently, to its fringe (Brookfield *et al.*, 1991). *Melayu baru*, in contrast, is not simply a matter of Malays being in the city; it is about urban(e) ways of 'being Malay'. Apart from working practices, globally-appropriate modes of urban(e) self-realisation span domains as diverse as business and entrepreneurship, driving, dressing, comportment, consumption and, as an important sub-set of the latter, eating (Sloane, 1999). To the extent that new global (or globally-appropriate) modes of being Malay in the city accommodate 'other' social groups – and to the extent that these reciprocate – new Malays make, as well as use, space for a new urbanity.

It is those spaces and social practices deemed out of place in the globalising city that make known the boundaries of what is acceptable or appropriate. Transgression is 'diagnostic' of the normative cityscape (Cresswell, 2000). The *kampung*, as we have noted, is important here.

Kampung has, in fact, long denoted more than merely an undesirable space in, or feature of, the Malaysian urban landscape; it has also served as a shorthand for those attributes, attitudes and modes of conduct deemed unsuitable for urban(e) life and for citizens of a would-be 'fully developed' nation (Mahathir, 1993). For city authorities, '*kampung*' – and especially '*kampung setinggan*' ('squatter settlement') – has come to signify anti-urbanity (Bunnell, 2002c). City *kampungs* have been rendered problematic by a diversity of 'experts' ranging from state policy-makers and international agencies to religious and academic authorities. Academic reports have long alluded to a widespread 'primitive level' of rubbish disposal (Pirie, 1976: 56). Yet what Phang Siew Nooi and others termed the 'dark side' of the squatter settlement included a wide range of supposedly 'inadequate' or 'improper' living conditions (Phang *et al.*, 1996: 133). While the increased urgency of squatter resettlement and the eradication of their *kampungs* was partly a response to rising demand for city land, the problematisation of '*kampung*' in the Kuala Lumpur cityscape in the 1990s remained a matter of aesthetic and moral concern as well as of economic calculation. The official goal of making Kuala Lumpur into a 'squatter-free city' was therefore bound up with broader governmental attempts to realise a new kind of city-dweller through appropriate urbanisation. City Hall's stated policy in the 1990s was 'to resettle the squatters into planned residential environment (*sic*) with all modern amenities and facilities' (Mokhtar, 1993: 17).

As in many other cities, high-rise flats in particular were technologies of residential modernisation. On-going amendments to City Hall housing estate and building design clung to a governmental faith that future conduct could be finely tuned in desired directions. The flats being built in Kuala Lumpur in the 1990s were of a 'third-generation' design. Referring to the process of squatter 'modernisation', the Deputy-Director in City Hall's Economic Planning and Social Amenity Department suggested that 'slowly their [squatter's] attitudes are changed in the flats'. Yet the process clearly was all-too-slow. The same informant also complained of 'problems caused by undisciplined people who still live like they did in the *kampung*'.

Anti-urban conduct among resettled squatters in high-rise blocks was also 'explained' in terms of a *kampung* (ab)use of space to be corrected through environmental education. In May 1997, the tragic consequences of a supposedly 'kampung' act elevated the Putra Ria 'low-medium' cost public flats to a symbolic centrality in authoritative evaluations of urban (mis)conduct. A twenty-seven-year-old technical assistant was killed by a brick thrown from block 94 of the Jalan Bangsar flats. Around one hundred of the Putra Ria flat units were occupied by former residents of the adjacent Kampung Haji Abdullah Hukom squatter settlement and they were concentrated in block 94. A front page report of the incident in the English language *Malay Mail* newspaper carried the subheading 'kampung habits die hard' and noted that:

> Although it has been nearly a year since they were relocated from Kampung Abdullah Hukom to the new flats, the residents never really discarded their habit of indiscriminate rubbish dumping.
>
> (*Malay Mail*, 1997a: 1)

In the rash of proposed governmental responses – which included more stringent enforcement of existing housing regulations, the introduction of high-tech surveillance technologies and even the decidedly low-tech 'binocular watch' (*Malay Mail*, 1997b) of errant tenants by City Hall – one is perhaps deserving of elaboration. The father of the block 94 victim was reported as advocating 'educational programmes ... particularly for squatters or new occupants of high-rise before they are relocated' so that they can 'adapt to the new surroundings' (*Malay Mail*, 1997c: 2). City dwellers were thus shown to have interiorised conceptions of the necessity of new forms of (self-)government as a response to 'kampung ways'. 'Kampung folk' strived to realise themselves environmentally within urban(e) limits.

It was not only in the squatter *kampung* or modern resettlement estate that transgressive acts made known urban(e) conduct. While as we have seen, the Kuala Lumpur City Centre and other putatively world class infrastructure showcased what Malaysia(ns) 'can do', newspaper reports repeatedly juxtaposed this alongside what city dwellers actually did (or failed to do). In a high-profile case in 1997, Prime Minister Mahathir angrily rebuked Malaysians after being 'ashamed' by comments by his Pakistani counterpart, Nawaz Shariff, about the rubbish in the Bukit Bintang area of Kuala Lumpur. The 'culture of rubbish' (Nelson, 1997) that had 'turned Jalan Bukit Bintang into a dump' (Shareem, 1997a: 2) was considered to have impacted adversely upon the image of the city and the nation. Such behaviour was bad, in part, because it was bad for business. The visibility of the Petronas Towers – in the city and internationally – was not sufficient to ensure that Malaysia was seen to be 'modern' and 'investible'. More significantly, perhaps, the visibility of problematic conduct in this case and others like it carried 'cultural' lessons for globally-appropriate conduct. Behaving 'better' entailed individual and collective environmental responsibility as opposed to an anti-urbanity where city space is considered to 'belong to nobody' (Nelson, 1997).

New forms of environmental citizenship, however, were not necessarily considered achievable merely through urban(e) (self-)improvement. Problematic social conduct appeared to have arisen as part of the process of urban development itself. According to the *Seventh Malaysia Plan*:

> Rapid industrialisation and the consequential rise in urbanisation and rural-urban migration have resulted in an increasing occurrence of negative social behaviour. Social problems such as drug addiction, child abuse, loafing, juvenile delinquencies, unhealthy lifestyles and strains on the family are beginning to emerge.
>
> (Malaysia, 1996: 27)

'Social ills' were made known in the 1990s by diverse socio-cultural and political authorities, eliciting a range of contesting explanatory accounts. In seeking to explain one of the most frequently-cited social ills among (especially Malay) youths, *lepak* ('loafing' or 'loitering in malls'), one political commentator noted that, 'there is no way for them to expend their energy – no parks, fields or other facilities' (Syed Husin Ali, cited in *Sunday Star*, 1997: 17). An earlier letter to the *New Straits Times* lamented how, 'our younger generation spends its time in shopping complexes mimicking the free-living style of the West' (Shaharuddin, 1991: 11). Both of these critical explanations diagnosed problematic conduct as being a result of – rather than resistance to or ignorance of – a globally oriented phase of urban development: whether in terms of the diminution of green space for global project(ion)s; or else as a result of the rise of 'Western' architectures of indolence. The evolutionary underpinnings of post-colonial state conceptions of development (Mahathir, 1970) arguably foster an expectation of such social ills as an enduring feature of urbanisation. The disproportionately high incidence of social ills among Malays, according to such systems of evaluation, signalled less an unwillingness to realise themselves appropriately than an inability to adapt to urban(e) life.

KLCC was also reported as a site at which opportunities for appropriate Malay self-realisation were more wilfully spurned. We have seen already how the project was positioned as an exemplary sight for *Melayu baru*. Yet KLCC was also intended as a site at and through which more Malays could become new Malays through participation in a world class construction. The project featured a 'technology-transfer scheme' in which every expatriate 'expert' had a Malaysian (and preferably Malay) under-study.[32] The Minister of Education publicly reminded KLCC of its 'social responsibility' to transfer skills and technology to *bumiputera* (cited in *New Straits Times*, 1995: 2), and expatriates spoke of a pressure to demonstrate 'local participation'.[33] The expansion of Malay capabilities and experience which has formed a prominent part of nation building since NEP was thus facilitated by participation in the physical task of building the capital city and the nation. However, the technology-transfer scheme was reported to have experienced difficulties, partly because of labour shortages in Malaysia,[34] but also because of a supposed Malay preference for 'easy' public sector employment and/or a propensity to 'work in the background' when they did venture into private sector employment (*New Straits Times*, 1995: 2). Some expatriate professionals internalised these post-colonial discourses of Malay unsuitability for, or unwillingness to engage in, competitive employment;[35] others complained of high turn-over of staff, particularly in middle-level positions.[36]

There were other ways in which the construction of KLCC was not a simple matter of existing urban practices giving way to 'global' standards. The stalls on Jalan Pinang where workers ate breakfast before starting work – and where I chose to kill time waiting for a 'real' interview, rather than facing 'the jam' – were alive with rumours about Chinese-subcontractors who had abandoned work on the lift-shaft of Tower Two after a number of fatal accidents. Seven lives were lost during the construction, six of which occurred at the South Korean-led tower (Oh, 1996). Stories (and storeys) of hauntings exemplify the presence of 'traditional' knowledges in what was clearly intended as an architectural statement of Malaysia's entry to the modern world. Another tale, in this case related to me by an expatriate working at the KLCC site, concerned an *imam* (Muslim 'prayer leader') who was brought from the Islamic University to send two evil spirits to the Genting Highlands.[37] The construction of other large-scale modern projects was also accompanied by tales of *feng shui* ceremonies and other traditionalist practices. The prevalence of such tales makes them less the traditional 'other' to an advancing global urbanity than a constitutive part of modernity's ongoing reworking in the city.

A new urbanity, then, was not reducible to Kuala Lumpur taking up its place in an undifferentiated and already-existing 'global'. Place itself mattered in the forging of an urban(e) social code (and, we might add, in the remaking of global modernities). In part, this may be understood in terms of the enfolding of in situ everyday practices into the 'world class' remaking of urban(e) space. However, 1990s Malaysia also bore witness to official valorisation of supposedly local conceptions and uses of space. In response to an apparent 'lack of community' in urban areas, Prime Minister Mahathir himself called for Malays(ians) migrating from rural to urban areas 'to practice their culture and lifestyle in their new surroundings' (*New Straits Times*, 1996b: 8). One of the newspaper articles on the Bukit Bintang incident alluded to above contrasted the 'culture of rubbish' in the city with the cleanliness of villages and their strong sense of collective ownership. Sociologist Norani Othman is cited as prescribing *kampung* as a model of collective environmental responsibility to be translated to the urban context. '*Kampung*' here is sanitised as part of a 'traditionalism' associated with the new middle classes (Kahn, 1992). Ironically, the valorisation of *kampung* as a model of urban(e) conduct was taking place while the *kampung setinggan* associated with urban poor disappeared from the urban landscape at an accelerated rate (Bunnell, 2002c). New, or newly-valorised, 'local' conceptions of space and society informed landscaped ideals of global urbanity.

This chapter has considered the global reorientation of Kuala Lumpur in the 1990s focusing on the new Kuala Lumpur City Centre (KLCC) project. Much of the chapter has been concerned with the symbolic meanings of the project and its iconic Petronas Twin Towers in particular. Like skyscrapers in other parts of the Asia-Pacific region in the 1990s, the Petronas Towers were intended to perform a civic role, placing the city and nation on 'world maps'. The work of making the towers stand for the nation was premised on a centralisation of authority over the city and a concentration of investment at a specific urban site. Yet I have also shown how 'global' imaginative reorientation was intended to have internal or domestic effects. The towers were bound up with new 'world class' ways of seeing and being Malay(sian). Diverse and dynamic official symbolic meanings, however, were subject to popular reworking as the Petronas Towers came into view(s) in the city. The building assumed a symbolic centrality in everyday dissatisfaction as well as in more explicitly 'political' discontent generated by the globalising phase of urban-centred development.

As we have seen, however, KLCC was not merely symbolic or representative of the unfolding of a 'global' round of geo-historical development. We considered the active role of the site/sight in shaping new geographical subjectivities. This forms part of broader conceptions of the constitutive effects of space. KLCC is thus understood in terms of a dynamic normative cityscape. Socio-spatial codes for appropriate, urban(e) modes of individual and collective conduct were revealed through transgression. In this way, transgressive anti-urban(e) acts in Kuala Lumpur made known an emergent new global urbanity. For those (new) Malays and other Malaysians who derived new skills and perspectives from global megaprojects such as KLCC, the Multimedia Super Corridor (MSC) – and its two new 'intelligent' cities in particular – provided an opportunity to realise norms and forms of Malaysian urbanity. It is to Putrajaya and Cyberjaya, the MSC cities south of 'KL', to which we now head.

5 Putrajaya and Cyberjaya

Intelligent cities, intelligent citizens

Phase 1 of the KLCC project in Kuala Lumpur coincided with a growing federal government emphasis on 'high-tech'. We have already noted how KLCC was incorporated into the Multimedia Super Corridor on account of its supposedly 'intelligent' features. Similarly, plans for the new Federal Government Administrative Centre, Putrajaya, roughly mid-way along the new southward-extending corridor at Perang Besar, came to emphasise high-tech aspects of living and working as 'the first intelligent garden city in Malaysia' (Putrajaya Holdings, 1997a: 2). The planning and development of Putrajaya, like KLCC, pre-dated MSC. A preliminary study carried out by the Economic Planning Unit of the Prime Minister's Department as early as 1991–2 concluded that relocation would be essential. This was said to be primarily in view of projections that civil service requirements would exceed Kuala Lumpur's capacity by 2010 (HG Asia, 1997a), but was also motivated by problems of traffic congestion and increasing travel times in the Federal capital.[1] Relocation to Perang Besar was justified by apparent efficiency gains from allowing government departments scattered across Kuala Lumpur to be housed in one location. Putrajaya, in fact, also had a long association with 'the most sophisticated and advanced information technology' (Jaafar, 1995: 5).[2] However, it is certainly possible to identify an 'intelligent' turn in the mid-1990s which reflected and fostered a technological utopian strand of state development. Cyberjaya, a 'model' cyber-city, was initiated as part of the MSC.

The Multimedia Super Corridor and its two intelligent cities in particular, were central to new applications of technology making known a 'multimedia utopia' (Multimedia Development Corporation, 1997a: 4). Seven primary areas of information technology were initially identified for testing in Putrajaya, Cyberjaya and neighbouring sites in the corridor (Multimedia Development Corporation, 1996b). 'Electronic government' and 'Telemedicine' referred to the employment of information and multimedia technology in government administration and Malaysia's healthcare system respectively. The 'Multipurpose Smart Card' would be developed to combine the existing national identity card with other government and private plastic transactions. 'Smart Schools' connected to the Internet, would build the necessary skill base for the information economy and society. Their initial development was scheduled to take place in 'Telesuburbs' to the south-east of Putrajaya (see Figure 5.1). Another Flagship Application, 'R&D Clusters', was intended to forge collaborative links between corporations and universities and these were initially to be tested in a Research and Development Centre. 'Worldwide Manufacturing Webs' and 'Borderless Marketing' sought to harness new technologies for regional manufacturing operations and electronic commerce (Multimedia Development Corporation, 1996c). Together the Flagship Applications were intended as the first seven steps to an 'intelligent' future in Malaysia.

Intelligent futures included, but also extended beyond, the merely technological or infrastructural. On the one hand, the new corridor was clearly the focus for massive planned transport and information infrastructure development. In addition to the digital optical fibre which formed the electronic 'backbone' of the MSC, integration of the new southern corridor with the existing Klang Valley urban region was to be facilitated by an Express Rail Link (ERL) and a Dedicated Highway to the airport. In focusing on development at MSC's intelligent cities, therefore, this chapter represents a move southward along the imagined transect of 'Route 2020' (Figure 5.1).[3] On the other hand, the way forward to an intelligent 2020 was bound up with the imaginative (re)construction of Malaysian citizen-subjects as much as with the physical construction of new high-tech spaces. One of the frequently-cited reasons for the building of Putrajaya was 'to start fresh on a new and large enough site to establish a planned city unimpeded by existing development' (Azizan, 1997: 2). For Malaysian planners and architects exposed to the 'world class' development as part of Kuala Lumpur City Centre Sdn Bhd – the project managers for KLCC who now came to play a leading role in Putrajaya – Perang Besar was a *tabula rasa* for a new national urbanity. In the context of the high-tech push into an information economy and society, this governmental challenge became one of seeking to enable would-be citizens of the MSC cities to realise themselves in intelligent ways.

The remainder of this chapter runs as follows. In the next section, I consider the discursive and material construction of MSC's intelligent cities. Putrajaya and Cyberjaya are shown to have been bound up with the imagining of new modes of work deemed appropriate for a world of high-tech. In the case of Putrajaya, this entailed both the electronic reworking of federal government administrative practices, and also a broader national technological shift as business and citizens would increasingly interact with 'electronic government'. Cyberjaya was envisaged as a technopole where world class high-tech companies would foment a culture of innovation in a wired city environment. I also consider a shift of emphasis from infrastructure to human resources in the state project of high-tech development. The active wooing of Malaysian expatriates and

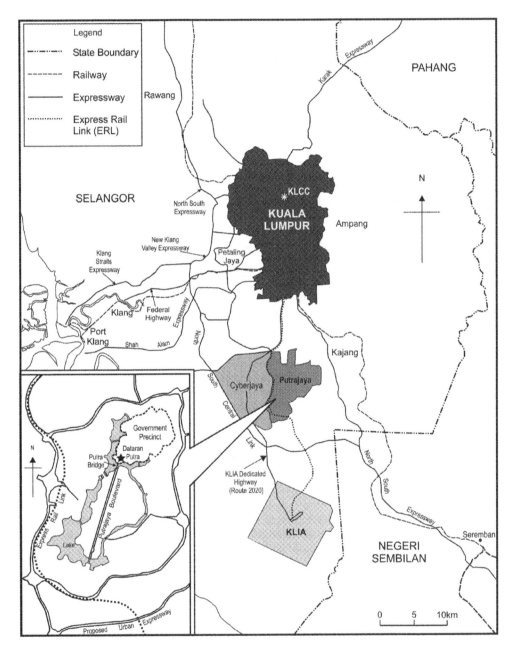

Figure 5.1 Kuala Lumpur Metropolitan Area.

other 'knowledge workers' from overseas, I contend, suggested potentially far-reaching implications for dominant post-colonial state (re)scriptings of national identity. The second section of this chapter considers the self-conscious envisioning of a putatively Malaysian conception of intelligent futures which extended beyond the strictly technological. Information society and economy was to be *led* (rhetorically at least) in desired 'national' directions which were not reducible to a replication of technologically-advanced socio-cultural formations elsewhere. Rather, intelligent national futures were imagined in part through the incorporation of supposedly 'indigenous' knowledges, notably from Islamic planning and *kampung* ('village') forms and norms.

In the third section of the chapter, I consider in more detail how intelligent city landscapes were imagined to be bound up in the realisation of intelligent citizens. Three dimensions of landscaped (self-)government are considered in turn: the first is *cerebral* or how individuals were to realise themselves innovatively in wired MSC urban environments in relation to a range of discourses of appropriate work (and non-work); the second is *social* or how forms of interaction and community were considered essential to individual and collective well-being; and the third is *environmental*, or how physical and moral health was to be promoted through active use of 'natural' environments. While none of these bio-political ideals, of course, is unique to the MSC, the possibility for their realisation in and through intelligent city landscapes is understood to make MSC a distinct space of (self-)government. In the fourth and final section of the chapter, however, I consider the role of MSC landscapes in the construction of nation identity. The symbolic national centrality of Putrajaya, in particular, was presented through design features and symbolism associated with other capitol complexes. MSC's role in the imaginative work of national landscaping was performed through a succession of high-profile, spectacular media events. Possibilities for an extension of MSC innovations across the national territory were imagined in relation to discourses positing an 'end of geography' associated with new (and especially information) technologies. More conventional technologies of exhibition and display allowed citizens to see themselves as part of intelligent *national* futures.

High-tech spaces and subjects

MSC was initiated to turn a 'greenfield site' into a high-tech space for suitably high-tech citizens and foreign 'knowledge workers'. In the case of Putrajaya – which, as we have noted, pre-dated the MSC – this meant specifically *national* citizen-subjects. Putrajaya was said to be a 'pioneer test-bed' for Electronic Government.[4] The expressed intention was not so much to upgrade existing administrative practices through computerisation as to 'employ multimedia technologies to re-invent the way the government operates' (Multimedia Development Corporation, 1996b: 7). According to the Deputy Director General of the Malaysian Institute of Microelectronic Systems (MIMOS), Dr Mohamed Arif Nun,[5] Putrajaya would be:

> the nerve centre, the information command and control centre to lead and steer the nation towards the achievement of the nine challenges in the Vision 2020, to enable Malaysia to compete successfully in the digital economy and eventually to lead in transforming the nation into an information rich and knowledge-intensive society.
>
> (Mohamed Arif, 1996: 1)

Such leadership was to be achieved through the creation of 'a multimedia, networked paperless environment that would link government agencies, businesses and citizens' (Professor Datuk Anuar, cited in Asohan, 1997: 45).[6] New technical practices and competencies in the federal administrative centre, in other words, would be extended to business and to individual citizens through their dealings with the government. The new administrative centre could thus be seen as a governmental programme in which the conduct and capacities of citizens were to be augmented and improved as part of a move to greater 'national competitiveness' (Mohamed Arif, 1996).[7] 'E-government' meant a technological shift which was understood to be cultural as much as infrastructural.

The initial development of MSC, however, emphasised infrastructural connectivity as the chief attraction for 'world class' company investment and foreign knowledge workers. Most prominent of all were boasts about information infrastructure, not least the development of a 'high-speed single backbone integrated telecommunication network to cater for advanced value-added telecommunications services and multimedia services' (Mohamed Arif Nun, cited in Corey, 2000: 139).[8] Yet elaboration of existing and planned investment in transport links was also highly prominent in the official literature. There are some important points to be made here about geographies of information society and economy. While the uniform and simultaneous availability of information made possible by information and communication technologies (ICTs) has been understood to diminish the significance of physical location, notions of the 'end of geography' have been debunked in at least three ways (see Graham, 1998). First, information economy and society is supported by complex infrastructure in 'real' places, including fibre optic cable as much as more visible technological investment (Graham and Marvin, 1996). Second, rather than negating the need for physical proximity, the most technologically sophisticated economic activities are frequently characterised by the continued or even heightened importance of face-to-face communication (Thrift, 1996a). And third, while at one level eroding the significance of spatial barriers through, for example, a cheapening of transmission and telecommunications costs, information technology may also augment sensitivity to other aspects of locational variation (Goddard and Richardson,

1996). Physical as well as electronic connectivity was thus essential for making MSC a viable locale for high-tech investment.

The locational appeal of MSC high-tech space was further enhanced by a range of 'soft' incentives and institutional arrangements. A Bill of Guarantees entitled MSC-status companies to: unrestricted employment of local and foreign knowledge workers; exemption from local ownership requirements; and the freedom to source capital and borrow funds globally (Multimedia Development Corporation, 1996c). While none of these incentives was without precedent in Malaysia, their combination in the Bill of Guarantees was initially exclusive to the 50 by 15 km MSC zone. More novel or specific to the 'high-tech' nature of the economy envisioned for MSC was a series of new legal provisions. In addition to the Multimedia Convergence Act, which sought to create an up-to-date communications framework, Malaysian Parliament passed an initial batch of so-called 'Cyberlaws' in 1997: a Digital Signature Bill to allow electronic signatures to be used instead of hand-written ones in legal and business transactions; a Copyright (Amendment) Bill giving intellectual protection through the on-line registration of works, licensing and royalty collection; a Computer Crimes Bill defining illegal access; and a Telemedicine Bill providing insurance for medical practitioners offering services from remote locations (Lim, 1997a). Together with the provision of telecommunications and logistics infrastructure, these new policies and cyberlaws for MSC enabled its proponents to gloss 'the perfect Multimedia Environment' (Multimedia Development Corporation, 1997a: np).

Shaping and legitimising this high-tech environmental transformation were two new institutions. One was the Multimedia Development Corporation (MDC), the lead agency in the management of MSC. Registered under the Companies Act, MDC functioned as a private sector entity with all the efficiency and productivity gains that this is meant to imply.[9] MDC was envisaged as a 'unique, performance-oriented, client-focused agency', a 'one-stop super shop' which negated the need for foreign companies to deal with multiple government agencies (Multimedia Development Corporation, 1996a: 8). As MDC's own marketing literature put it: 'In working with companies setting up operations in the MSC, the MDC will serve as promoter and facilitator' (*ibid.* 8). Dr Mohamed Arif Nun was seconded from MIMOS to become Chief Operating Officer of MDC in November 1996 (Teoh, 1997). A second body, the MSC's International Advisory Panel (IAP), also signalled responsiveness to transnational corporate investors. The panel assembled high-tech 'luminaries' such as IBM's Louis Gerstner, James Barksdale from Netscape, Eckhard Pfeiffer from Compaq, Noboru Miyawaki from Nippon Telegraph and Telephone as well as Microsoft's Bill Gates. The recruitment of Gates in particular, was hailed as a public relations coup for Prime Minister Mahathir (Johnstone, 1997). Yet the panel was clearly also intended as a way of convincing the 'Who's Who in the IT Hall of Fame' (Shareem, 1997b: 2) that Malaysia should be their primary investment location in Asia by allowing them some say over the development of the(ir) MSC test-bed.[10]

In addition to inviting the 'great minds' from Silicon Valley and elsewhere (Hutnyk, 1999: 315) to see MSC as their base in Asia, Mahathir took the concept to potential investors overseas. While emphasising MSC's 'intelligent', financial and institutional features, attempts to secure investment in information and multimedia industries also discursively constructed the corridor (and Malaysia) as an environment in which Asian cultures 'come together' due to the geographical location and multi-ethnic composition of Malaysian society. A speech in Los Angeles evidenced what might be understood as Mahathir's 'multicultural marketing':

> The Malaysians are made up of people of Malay, Indonesian, Indian and Chinese origin. We are only a few hours flight from the major Asian capitals. We have language skills and cultural knowledge that can be very helpful. Most people speak English as well as one or more languages such as different Chinese or Indian dialects, or Malay. ... Malaysia will be a highly efficient and effective hub for the region.
>
> (Mahathir, 1997b)

Thus, while not possessing either the labour pool or potential market size of Indonesia, China or India, Malaysia was imagined to have cultural resources that would enable easy access to each of these larger economies. This positioned MSC favourably in relation to 'hub' function competitors that were equally if not more attractive in terms of other locational criteria. Malaysia's 'multicultural edge' (*New Perspectives Quarterly*, 1997: 5) thus referred to a competitive advantage supposedly born of multicultural diversity (see Bunnell, 2002d). Yet Bill Gates' eventual decision to establish his 'Asian' Microsoft operation in Hyderabad, India, confirmed a heightened regional competition for 'world class' high-tech companies (see Hutnyk, 1999).

Apart from being promoted as a would-be regional hub for international corporations, Cyberjaya in particular was presented as a model locale for the formation of innovative Malaysian companies. The city was intended not merely as an industrial park, but rather as what has been termed elsewhere, a 'technopole' (Castells and Hall, 1994). The putative difference is that while the former merely provides investment in, and employment using, existing industrial techniques and technology, the latter is concerned with research and development – the creation of *new* technology. Undoubtedly the most celebrated technopole is Silicon Valley in the northwest quarter of Santa Clara County in Southern California. At one level, Cyberjaya may be understood in terms of a long and expanding list of attempts to replicate Silicon Valley's 'technical virtuosity and economic dynamism' (Winner, 1992: 32). Certainly, the oft-repeated aim in the planning of Cyberjaya was to produce an environment which was 'conducive for creativity' (Singh,

1997a: 1). The initial phase of the development began in a so-called Flagship Zone covering 2,800 of the 7,000 or so hectares of the greater Cyberjaya area. The 62-member team led by the director of the Federal Department of Town and Country Planning, which was responsible for the physical development plan, included staff from a number of Japanese corporations, including the Mitsubishi Research Institute and NT&T Corporation. Japanese corporate influence on the development of MSC, therefore, extended beyond the role of Kenichi Ohmae, Mahathir's multimedia adviser (see Chapter 3).

Cyberjaya planners visited a number of innovative technology environments: not only Silicon Valley and the Technopolis projects in Japan, but also Sophia Antipolis in France and Bangalore, India. One of the conditions common to existing technopoles was the proximity of a university or a similar research-based institution. The role of such institutions in technology generation elsewhere was recognised in plans for the MSC, and the Cyberjaya Flagship Zone included a Multimedia University aiming to 'enhance creative dynamics between research and industry' (Multimedia Development Corporation, 1997b: 11). This, in turn, related to MSC's R&D Cluster Application which aimed, 'to promote the development of next-generation multimedia technologies by forging collaborative R&D efforts among leading-edge corporations, public research institutions, and universities' (Multimedia Development Corporation, 1996a: 23).[11] That the MSC was envisaged as an advanced technological milieu – rather than a manufacturing, industrial park or free trade zone[12] – was further evidenced from the guidelines for achieving MSC status. A company seeking MSC status and eligibility for MSC incentives had to: be deemed 'a provider or a heavy user of multimedia products and services' and; 'employ a substantial number of knowledge workers' (*ibid*. 12). The intention was that foreign multimedia companies would foment a culture of innovation in which Malaysians could initially participate and to which they would ultimately contribute.[13] In response, no doubt, to a belief that local companies would begin as very small concerns, MSC also included plans for a 'Cybervillage' south of Putrajaya, 'the hub for local small and medium-scale enterprises (SMEs)' (Zainuddin, 1997: 3). MSC planners were clearly cognisant of the experiences of would-be high-tech others.

In keeping with high-tech discourses and practices elsewhere, emphasis gradually shifted from the attraction of financial investment to human resources. If NEP had prompted a generation of middle-class, 'non-Malay' Malaysians to seek education, employment and even citizenship overseas, new techno-economic imaginings began to fuel attempts to bring skilled professionals 'home' (See, 1997). As many as 40,000 Malaysian professionals migrated to the USA, Canada, Australia and New Zealand between 1983 and 1990 (Pillai, 1992). Many more studied overseas and stayed on to work, particularly in city-state neighbour and high-tech rival Singapore (Hing, 2000). Attempts to attract back overseas nationals which accompanied the shift in economic emphasis in the early 1990s gained momentum after the initiation of the MSC project. In part, this was stimulated by state recognition of the key role of South Korean and Taiwanese return migrants to their respective high-tech economies (see Pereira, 1997). MSC appeared to require skilled Malaysians and this raised issues about the kinds of social and political (as well as economic and technological) conditions necessary to 'woo' such people back. One newspaper article called for the promotion of 'the Malaysian version of the American Dream – that with brains and hard work, success is within everyone's reach' (See, 1997: 5). If the globalisation and regionalisation of the economy had initiated a multicultural state rescripting of national identity, this proposed 'American dream' necessitated a shift in state ways of seeing and differentiating its population from ethnicity to skills. Did state-sponsored attempts to woo back skilled non-Malay subjects suggest such a fundamental shift? Or was this, rather, a temporary measure to fill a shortfall in skilled labour until such a time as demand could be generated through 'indigenous' human resources?

Irrespective of the cultural politics of recruitment and repatriation, suitable high-tech subjects implied more than mere possession of *existing* skills. Rather, the prized citizen-subject (or 'knowledge working' migrant) was one able to adapt to and even generate future technological transformation. Appropriate conduct, in other words, was not reducible to adeptness in the use of 'Information Age tools and technology' (Multimedia Development Corporation, 1996b: 15). It was citizens' attitudes to learning and acquisition of knowledge which were increasingly understood as primary determinants of a successful realisation of high-tech futures. In Malaysia, as elsewhere, a globalising world was depicted as one in which intensified interconnectedness and the speed-up of transactions through technological change defined a period of massive uncertainty (see du Gay, 1996). In this context, 'entrepreneurship', 'enterprise', 'creativity' and 'innovation' were offered as solutions to the 'problems' of globalisation. Thus, in formal educational terms, so-called Smart Schools in the MSC were intended to 'create a new generation of Malaysians – Malaysians who are more creative and innovative in their thinking, adept with new technologies, and able to manage the information explosion' (Dato' Sri Mohd Najib Tun Haji Abdul Razak, cited in Ministry of Education, 1997). Malaysians were also compelled to think 'beyond the classroom' – whether as public sector workers appropriating new technology for more efficient and responsive administration or as innovators developing new technologies and applications in Cyberjaya's incubator centre – to realise the full potential of information society and economy.[14] Commitment to on-going learning and knowledge acquisition thus became a personal and national responsibility. Individuals were compelled to continually upgrade and adapt themselves for new forms of economy. In this way, learning citizen-subjects would eventually evolve to imagined territories of technological leadership currently occupied by 'Homo Silicon Valleycus' (see Thrift, 2000a: 688).

Intelligent national futures

There are two interrelated reasons why the imagined spaces and subjects of high-tech MSC futures were not reducible to mere replication of Silicon Valley's putative evolutionary lead. The first concerns MSC's relation to broader self-conscious attempts to articulate an ostensibly *national* conception of information society and economy. A range of authorities in Malaysia – socio-cultural and political as well as technological – grappled with what might or should constitute a specifically 'Malaysian' intelligent future. A second reason was precisely because such national ideals extended beyond the technological. The 'intelligent' was not restricted to the realm of wires and fibres but, understood normatively, came to connote an idealised national society and means to achieving it. In this section, I detail the imaginative work of state actors and other authorities in articulating intelligent Malaysian futures in the MSC.

MSC discourse made known an intelligent Malaysia which was both characterised by, and to be realised through, the extensive use of information technology. As depicted in official charts, MSC did not merely constitute the first phase of a route to a 'fully developed' 2020; it also served to set the trajectory for subsequent phases of national development which would eventually catalyse an explosive 'leapfrog into leadership in the Information Age' (Multimedia Development Corporation, 1997a: 10). The trope of 'leadership' – noted already for the development of Putrajaya – was significant, implying a conception of Malaysia(ns) actively defining high-tech times, new developmental goals and means to achieving them. The imaginative charting of intelligent Malaysian futures thus extended beyond conventional routes of/to development. One diagram produced by the Multimedia Development Corporation presented Malaysia's developmental 'leapfrog' as an EXPLOSION where technological futures depart from charted developmental territory (*ibid.* 10).[15] 'Leadership' also defied technological determinist understandings which would have precluded possibilities for planning or intervention. The so-called Information Age was not imagined as an undifferentiated new technological epoch which Malaysia(ns) had to strive to enter or which would impact upon them. Rather, Malaysians were compelled to harness new technologies to construct their own high-tech futures – to *lead* technological development in desired directions.

Malaysia's route was, of course, to be led by MSC. We have seen in previous chapters how the project was officially conceptualised as a 'test-bed'. This related in part to a representational politics of 'containment': 'We don't want the multi-media (technology) to be applied nationwide lest there are adverse effects. If there is something adverse we can control it since we are limiting it to the MSC' (Mahathir, cited in Ashraf and Radzi, 1997: 3). Yet, in keeping with broader discourses of 'Asian' development, Mahathir also envisioned possibilities for leading technology in Asian/Malaysian ways. In a dialogue in Fukoka, Japan, Mahathir stressed the continuing importance of 'Asian values' in the 'digital century': 'if anything, they are more relevant for our men and women as they search for their own niche and identities in a borderless environment' (cited in Kaur, 1997b: 26). There were thus important continuities with geo-political discourses of Asia in official representations of MSC as a social and cultural (as well as technological) testing ground.

The means to an intelligent society tested in the MSC showed continuities with, as well as important differences from, previous planning in Malaysia. On the one hand, while incorporating the latest information and communications technologies, Putrajaya and Cyberjaya embodied a modernist belief that architectural forms and the design of urban space could foster a new and better Malaysian society (cf. Holston, 1989 on Brasilia). Even the means for achieving this drew upon structure planning and Garden City concepts which were both well established in Malaysia (and which, of course, had influential precedents in other national settings, particularly Britain).[16] On the other hand, plans for the intelligent MSC cities sought to define specifically national land use and landscape guidelines, elaborating desired planning outcomes in opposition to urban social problems in Malaysia and elsewhere, and formulating strategies to reduce such outcomes by drawing upon 'indigenous' knowledges and practices (Zainuddin, 1995). The key problematic was this: while MSC's political proponents strived to emulate Silicon Valley's technological lead, urban planners were among those authorities in Malaysia seeking *not* to replicate urban America's high divorce rates and incidence of family breakdown – what, in the case of Southern California, has been termed the 'dark side of the chip' (see Siegel and Markoff, 1985; and Fergusson, 1997). State planning discourse, was mindful of the apparent social death of once 'great' American cities (and declining Western civilisation more broadly):

> As cities in America grew, the society in them died a slow death. There is a missing ingredient in the planning and development of cities in America. For this nation [Malaysia] not to suffer the same fate of great American cities and more importantly, for society not to deteriorate as in societies of the west, that missing ingredient has to be found.
>
> (Zainuddin, 1995: 18)

According to Professor Zainuddin bin Muhammad, Director of the federal Department of Town and Country Planning, the missing ingredient was identified as 'the integration of spiritual and moral values into planning and development' in Malaysia (*ibid.* 18–19)

The Department of Town and Country Planning's so-called Total Planning Doctrine (TPD) elaborated and refined alongside the development of Putrajaya and Cyberjaya, established national principles for planned urban social health. In its initial manifestation, TPD consisted of a three part design ideology: 'Man and his Creator', 'Man and Man' and 'Man and Nature' (*sic*) (*ibid.*, 1995: 19).[17] It is perhaps symptomatic of broader shifts in national development discourse in the mid-1990s that later plans for Cyberjaya included a fourth element, 'Man and technology'. However, having considered the specifically technological underpinnings of MSC's would-be 'intelligent' futures, it is the other three (extra-technological) elements on which I will focus here.

According to Professor Zainuddin, the TPD arose from Prime Minister Mahathir's demand that the new federal government administrative centre be based on a notion of *bandar firdaus* (literally, 'heavenly city').[18] Having read the verses on *firdaus* – the Islamic view of paradise – in the Koran, Zainuddin not only formulated TPD as the essential series of relationships for 'civilisation', but also identified specific urban design features for Putrajaya such as the presence of clean water, gardens and the centrality of a mosque.[19] The foregrounding of issues of 'Man and his Creator' connected Putrajaya up with broader tensions between 'Islamic' and 'national' visions of social health. On the one hand, ostensibly 'Islamic' development served to appropriate – or in Professor Zainuddin's words, to 'shut down' – political critique of state development couched in Islamic terms. On the other hand, when applied to the new centre of federal government, such principles may be said to have symbolically marginalised non-Muslims from the post-colonial nation-state. Perhaps in recognition of this, proponents of the city stressed that Putrajaya was intended to 'satisfy the … spiritual needs' of followers of *all* religions practised in Malaysia (Azizan, 1997: 3). In addition to the promise of provision of spaces of worship for religions other than Islam, in interview, Professor Zainuddin emphasised histories of religious tolerance in Islamic cities. Besides, he suggested, TPD referred to 'Man and Creator', not 'Man and Allah which is Muslim-based'. Nonetheless, some official representations played down the Islamic framing even further. In advertisements for Putrajaya, allusions to 'Mother Nature' may be said to have collapsed the division between nature/environment and 'Creator'. It is also perhaps significant that, in the case of Cyberjaya, the term 'Man and his Creator' was frequently replaced altogether with the 'spiritual and health' aspects of development (Wan Mohamad, cited in Singh, 1997a: 1).[20]

The prominence of extra-economic aspects of national development in TPD – as in Vision 2020 – was partly a response to the perceived dark sides of Malaysia's own social transformation. We noted in the previous chapter perceptions of rising 'social ills' associated with urbanisation and modernisation. It was in part relations between 'Man and Creator' that appeared in need of attention here: 'religious innovation' among non-Muslims (Ackerman and Lee, 1988) as well as Muslims (Chandra, 1987) was reportedly a response to the spiritual alienation caused by rapid social and economic change. Yet previous critique of conditions in Kuala Lumpur had also highlighted problematic relations between 'Man and Man', notably in terms of a supposed lack of civic consciousness. TPD's 'Man and Nature' similarly sought to rectify existing urban problematisations, in this case an 'imbalanced' relationship between people and their environment both caused by, and resulting in, 'inappropriate' development. The planning guidelines for Putrajaya and Cyberjaya, therefore, emphasised the importance of preserving and/or working with the 'natural' environment. One of the chief aims of Cyberjaya was said to be to 'educate' developers to carry out their work in more 'sensitive' ways.[21] The city was to be developed in accordance with the existing topography which meant minimal cutting and filling and that hillsides over forty metres in elevation 'will be exempt from development and be designated as natural landscape' (Department of Town and Country Planning, 1997: 31).[22] Malaysia's intelligent cities thus both drew upon and extended 'planning philosophies ... adopted from the West' (Zainuddin, 1995: 17); yet *bandar firdaus* was also envisioned, in part, in opposition to the existing 'K-Hell' (*New Straits Times*, 1997b: 1).

Malaysia's intelligent cities also drew upon a rather different source for ideas on national development: the *kampung*. The *Landscape Master Plan for Cyberjaya* understood the *kampung* as an example of

> ... a landscape in which human settlement maintains a sustainable level of development, in balance with the natural environment. Resource use does not exceed the regenerative powers of the land.
> (Department of Town and Country Planning, 1997: 46)[23]

The practice of landscaping at Putrajaya was similarly said to be informed by '*kampung*' knowledges. Professor Zainuddin described how he devised a solution to providing clear water in the lake at Putrajaya – in line with the goal of becoming *bandar firdaus* – by thinking back to childhood *kampung* experiences in Kelantan. The fighting fish he caught in the *padi* fields, he recalled, could be found around a particular kind of weed where the water was very clear. As a 'natural filter' for the dirt, this weed was introduced to Putrajaya lake and wetlands. Such 'indigenous' innovation forms part of a broader valorisation of '*kampung*' environmental understandings and approaches (see also Sloane, 1999). One coffee table book produced a decade earlier, Lim Jee Yuan's *The Malay House: Rediscovering Malaysia's Indigenous Shelter System*, had elaborated *kampung* in terms of a 'design-with-nature approach' (Lim, 1987: 68).[24] Dealing primarily with traditional Malay architectural forms, Lim focused on *kampung* housing, 'designed with a deep understanding and respect for nature' (*ibid.* 68).

The revalorisation of *kampung* in Malaysia has extended beyond the 'environmental' or landscaping aspects of national development. As we saw in the previous chapter, *kampung* traits and values have increasingly been understood as a remedy for supposedly undesirable social characteristics of modern life in Kuala Lumpur (see also Bunnell, 2002c). Lim Jee Yuan had depicted inappropriate modern development as a 'process of moral neutralisation' of previously harmonious communities (Lim, 1987: 144). The orientation of such writing, however, was not so much nostalgic as prescriptive: indigenous remedies for inappropriate modern development. Perhaps the convenient English translation of the word *kampung* as 'village' – denoting a physical or administrative area – misses a sense of the concept as a set of relations between people, a 'community' (see Shamsul, 1988).[25] The 'Kampung-minium' concept coined by another Lim – President of the Malaysian Institute of Architects (PAM), Jimmy C. S. Lim – had 'the "inter-dependent" lifestyle of the Kampong . . . the feeling of community intimacy' as one of it is principal objectives (Lim, 1993: 36; and see also Goh, 2002a). In 1993, his concept was presented to the federal government ministry responsible for the planning of the MSC cities (Cheah, 1993).[26]

The significance of *kampung* to discourses of 'intelligent' MSC futures may be summarised in four ways. First, like Islamic knowledges and ideals, *kampung* was incorporated into attempts to formulate an ostensibly national vision of development – and one that extended beyond the purely 'technological'. Second, while state actors and institutions were clearly important, authorities involved in defining intelligent national futures also included academics, the 'Malay cultural industry' (see Kahn, 1992) and design professionals in some cases only loosely, if at all, connected to the government. Third, this range of cultural authorities was both inclusive of non-Malay individuals and posited national futures which were ostensibly Malaysian rather than exclusively Malay/Muslim.[27] Fourth, *kampung* here was understood not merely as a space or way of designing space, but also as a set of relations. *Kampung* thus came to inform conceptions of how Malaysians should realise themselves individually and collectively in intelligent city spaces. Planning here, in other words, was conceived as a governmental rather than as a purely technical practice (see Mercer, 1997). It is to the 'government' of intelligent landscapes that I now turn.

The governmental work of intelligent landscapes

The MSC, as we have noted above, emerged as a space where diverse attempts to articulate national futures came together. The intelligent cities of Putrajaya and Cyberjaya in particular were designed for suitably intelligent citizens. However, appropriate national ways of living and working here were clearly not merely anticipated outcomes of particular architectural arrangements or environmental design. If, as Don Mitchell has suggested, landscape is something that 'does work', then this is not to imply simple spatial or environmental determinism (Mitchell, 2000). The idealised citizens of MSC texts and plans were those who would realise *themselves* in and through intelligent cityscapes. While Islamic, *kampung* and other knowledges made known appropriate conduct, the MSC cities were intended as assemblages of technologies to facilitate such (self-)government. Landscape was thus imagined to work in the conduct of intelligent conduct.

In this section, I outline three dimensions of would-be intelligent government in and through MSC landscapes.

The first dimension is *cerebral*. We have already seen how intelligent subjects were imagined as learning, creative and innovative. Cyberjaya in particular, was designed as an environment conducive for innovation and creativity: one in and through which foreign and Malaysian knowledge workers and their families could realise themselves in innovative ways. Significant here was the guarantee of no Internet censorship.[28] While clearly motivated by the demands of foreign inward investors – and serving to make Malaysia appear more 'information-friendly' than neighbouring, high-tech rival, Singapore (see Hiebert *et al.*, 1997) – the 'free flow' of information was also understood as essential for a place where 'creativity and innovation can thrive' (Mahathir, cited in Ramlan *et al.*, 1997: 1). As a technopole, Cyberjaya was intended to allow locally-sited firms to plug into and contribute to extensive inter- and especially intra-firm networks of innovation. Intelligent locality and extra-local interconnectivity, in other words, are inseparable. Beyond the firm, an environment of total connectivity was understood as a means of allowing workers and other residents to continually (re)educate and 'upgrade' themselves globally.

This raises the question of how 'free' citizens could be induced to realise themselves in 'learning' ways. Geographical research in other contexts has highlighted the range of objects, images and texts which shape subjectivities for 'new' forms of economy. Nigel Thrift, for example, has examined *Fast Magazine* as bound up with the realisation of 'fast' subjects (Thrift, 2000a). Similarly, in Malaysia, the launch of the MSC was associated with the emergence of new business magazines and programmes making known appropriate and desirable modes of entrepreneurial comportment and conduct. If it was the corporate elite of the International Advisory Panel (IAP) that gave the MSC 'world class' legitimacy, it was the home-grown figure of the Malaysian-made-good that demonstrated routes to new ways of being Malaysian.[29] While the mere physical design and construction of innovation-conducive city spaces would not be sufficient for innovative conduct, spatial arrangements were clearly considered constitutive technologies for learning selves. In the MSC cities, urban design, for example, was imagined to play an active symbolic role in relation to Islamic knowledges. Planning guidelines from the cities elaborate how 'symbols of knowledge' in the urban landscape stimulate consciousness of the Creator in turn compelling knowledge acquisition. Apart from the provision of sites of (self-)learning in the cities – Cyberjaya's Multimedia University and libraries as well as 'free' (that is, uncensored) Internet access – the presence of religious landscape artefacts was intended as a reminder of the duty of self-education and striving for knowledge (see Department of Town and Country Planning, 2001).

Such cerebral striving and the all-important goals of innovation and creativity were not imagined solely in terms of *individual* corporate ambition or religious duty. Cyberjaya planners in particular conceived of innovation as an outcome of human interaction. While we have noted the increased technological possibilities for such collective processes to occur at a distance, the sought after local 'buzz' (Hong, 1997a: 16) implied more than the sound of monitors connecting spatially dispersed individuals. Co-proximity and 'face to face' interaction remain important for the transfer of 'tacit' knowledge (Howells and Roberts, 2000). In keeping with would-be innovative locales elsewhere, Cyberjaya saw the establishment of 'incubators' for small firms to be housed together (Multimedia Development Corporation, 1997c). The wider goal was one of 'mutual enrichment' not only within and between firms, but also between Malaysians and foreign knowledge workers. A critical mass of companies and residents was thus considered a necessary precursor to 'the development of a highly competitive cluster of Malaysian multimedia and IT companies that will eventually become world class' (Mahathir, cited in Ramlan *et al.*, 1997: 1). This was itself premised on being able to attract existing companies away from sites in and around Kuala Lumpur with established infrastructure and social resources conducive for high-tech work.[30]

This connects with a second dimension: the *social*. In attending to relations between 'Man and Man', planners imagined particular forms of human interaction as a necessary ingredient for social health. As in previous planned residential developments, the 'neighbourhood' was the scale at which new community bonds would be fostered. One text on Putrajaya described how:

> Each of the neighbourhood units will be provided with facilities designed to promote increased social contacts and neighbourly interactions which sadly is (*sic*) rapidly eroding in our pursuit of material progress.
>
> (Azizan, 1997: 3)

Facilities such as community and neighbourhood 'focal points' (Putrajaya Corporation, 1997: 9), playgrounds, kindergartens and school complexes would thus result in 'a community way of life that encourages high moral values' (Shareem, 1997c: 16). This idealised moral community was rendered visible by looking both forward and back. On the one hand, the search for positive forms of social interaction was frequently articulated using Vision 2020 phraseology: Putrajaya would 'enhance the role of the nation in nurturing a caring, progressive society' (Putrajaya Holdings, 1997a: 4).[31] On the other hand, the way forward here was also made known by recovering supposedly pre-modern or 'traditional' values. Given the valorisation of 'indigenous' knowledges noted in the previous section of this chapter, it is no surprise that the decision to prohibit the construction of fencing between properties in Putrajaya was made in relation to the imagined 'open society' of 'past' *kampung* lifestyles (John, 2000: np). As in

Garden City experiments, then, imagined rural social health was to be combined with urban facilities and dynamism.

In these neo-traditional cities, the family was imagined as the collective foundation through which individuals could realise themselves socially. Cyberjaya plans featured a so-called 'neighbourhood planning theory' intended to 'promote a lifestyle of social interaction among all the residents of the community' (Department of Town and Country Planning, 1997: 39). According to this, 'community' would be the largest in a five-level scale of 'living units'. The other four, in descending order, were: neighbourhood, communal, housing group and, finally, family. Housing groups would comprise twenty to thirty families and feature 'small open spaces such as cul-de-sacs, secondary access roads and small squares are viewed as places for everyday contact between neighbours' (*ibid*. 39). Roughly 1,500 housing units were planned to comprise a neighbourhood, each with its own centre, school and neighbourhood park. One advertisement for Putrajaya depicted a family enjoying the great outdoors under the heading, 'the ideal place to work and raise a healthy, happy family' (Putrajaya Holdings, 1997b: 24). The father points to something out of view, satisfying his child's healthy curiosity and eagerness to learn – in line with knowledge-based living. Family was similarly valorised in the promotion of 'healthy' interaction in cyberspace. According to then Deputy Prime Minister, Anwar Ibrahim, 'The strongest "firewall" is the one which is built by oneself and family based on a system of values and religion' (cited in Hong, 1997b: 5).[32] Thus, the family and religion became known as semi-autonomous domains for the promotion of appropriate electronic learning and socialisation.

There was clearly a tension between ideals of fomenting social interaction through family and community, on the one hand, and the replication of leading edge innovative environments on the other. We have noted already how MSC planners appeared mindful of the social 'dark side' of technologically exemplary milieus such as Silicon Valley. Yet to what extent would a 'community life that encourages high moral values' (Shareem, 1997c: 16) appeal to a global knowledge working elite? There is a point to be made here, of course, about the different emphases involved in selling the MSC to international investors as opposed to domestic audiences. While addressing overseas audiences, MSC proponents stressed their liberalism. Following Dr Mahathir's visit to Hollywood – which included a meeting with the resident Malaysian community at a local mosque – IT adviser, Tengku Azzman Shariffadeen reassured *Asiaweek's* readers: 'I met a Malaysian working with [Steven] Spielberg who has a pony tail. I can be surrounded by such characters. They don't affect my values' (cited in Mitton, 1997: 25). Yet whether this apparently 'intelligent' Malaysian would come to feel 'at home' back in neo-traditional MSC cities was another matter. As suggested in the same *Asiaweek* article in which Azzman's comments were reported, '... Malaysian values could collide with the notoriously eccentric ways of computer nerds who will arrive in droves if the corridor takes off' (*ibid.* 25). A prior issue, of course, was simply whether such 'nerds' could be attracted to life in Cyberjaya in order to facilitate its 'take off' as an innovative milieu.

A third dimension, the *environmental*, does acknowledge and seek to respond to potential social problems of high-tech work. One of the reasons for building intelligent cities from scratch was the supposedly salubrious effect of sylvan landscapes. Attractive surroundings were not only considered necessary to attract 'cyber-experts' (with or without their families) in the first place, but were also understood to play an important role in augmenting individuals' work productivity. Experience of 'nature' would counter stresses from long hours of competitive employment particularly in what was seen as the 'artificiality' of a life of cerebral activity. This faith was evident from the *Landscape Master Plan for Cyberjaya*:

> In such an environment, where so much of a person's experience will be cerebral and based on conceptual rather than physical frameworks, an instinctive desire to have tangible experiences of physical reality, particularly nature, will arise. To relax from the mental strains of work and daily living, people will need the peaceful physical environment of natural surroundings and quiet outdoor spaces.
>
> (Department of Town and Country Planning, 1997: 47)

MSC city planners thus accorded the pervasive belief in the restorative power of 'natural surroundings' and 'outdoor spaces' a particularly high-tech urgency.

The understanding that a more 'balanced', harmonious relationship between 'Man and Nature' can realise a healthier population has underlain a long tradition of architects' and planners' Utopian 'visions of perfection' (Markus, 1985; see also Cherry, 1988): from Ebenezer Howard's Garden City concept to post-war British new towns (Hall, 1988). Malaysia's own early new towns followed British models[33] and, more recently, Bangi – a Malaysian new town southeast of Kuala Lumpur which opened in 1987 – was planned in a 'garden style with generous open space, playgrounds, lakes and recreational grounds' (Lee, 1987: 160). The MSC cities, to a large extent, followed these precedents. Forty per cent of Putrajaya was said to be 'natural' (Putrajaya Holdings, 1997a: 6),[34] a figure that included twelve parks and gardens (including a 203 hectare Metropolitan Park and a 63 hectare Urban Forest Park), a wetland sanctuary and botanical gardens (Putrajaya Corporation, 1997: 10). Similarly, the *Landscape Master Plan for Cyberjaya* allocated, 'in excess of 40% of the entire Cyberjaya project to greenery and open space' (Department of Town and Country Planning, 1997: 31). The city's 'Nature Zone' was described in terms of a therapeutic communion with nature: visitors would be able to 'enjoy the feeling of being in the woods, hearing the birds sing, watching fish and insects, relaxing to the music of streams ...' (*ibid.* 1997: 31).

Appropriate environmental conduct, however, was imagined to extend beyond merely looking or listening. Planners sought to induce active use of the environment and the 'outdoors' to promote physical and/or moral welfare. We saw in the previous chapter how various 'social ills' in Malaysia were attributed not so much to a lack of open space *per se* as to an excess of 'passive space'.[35] Thus, the MSC cities were intended to enable a broad range of leisure and recreational activities in the 'natural environment'. One of the stated objectives of Cyberjaya, for example, was to realise a 'landscape interactive lifestyle' (Department of Town and Country Planning, 1997: 46) through activities of 'right living' such as hiking, picnicking and nature study (cf. Matless, 1995).

Space, then, was understood to perform something in the realisation of each of these three dimensions of intelligent conduct. As we have noted, this is not to suggest a deterministic relation. Rather, understood governmentally, the MSC cities were intended as bio-political environments in and through which 'free' individuals could realise themselves appropriately. Few of the planning elements that were to comprise this environment were unprecedented but, when combined with the soft incentives of MSC's *Bill of Guarantees*, they formed a distinct space of government. As we will consider in the next chapter, there were distinct differences in the ability of people and places to realise themselves in suitably 'intelligent' ways, both *within* MSC as well as between this space of government and other parts of the national territory. Nonetheless, in political terms, MSC was also a powerful symbolic national landscape and it is to this that we now turn.

New technologies of nationhood

While elaborated as a 'test-bed' cut off from the rest of the national territory, MSC and its intelligent cities in particular played a powerful symbolic role in articulating Malaysian national identity. We have seen already how Putrajaya and Cyberjaya were bound up with attempts by a range of authorities – including, but not only, government actors – to construct an ostensibly *national* version of information society and economy. There was clearly also an important politics to (re)presenting the new cities as distinctly 'Malaysian'. In this section, we consider how urban design and landscape architecture, mediated to national audiences, became 'technologies of nationhood' (Harvey, 1996: 56). We also consider how new information and multimedia technologies were mobilised as part of this political process. The very global, 'information age' technologies often associated with the demise or 'end' of the nation-state (Ohmae, 1995) were thus mobilised in a state-led imaging of intelligent Malaysia.

The symbolic work of the new seat of Electronic Government, Putrajaya, incorporated planning imagery and vocabularies found in other 'capitol complexes' (see Vale, 1992). The metaphor of Putrajaya as the 'nerve centre of the nation' (Putrajaya Holdings, 1997c: 13) implied political power emanating from a central administrative hub. Not only was Putrajaya to be the 'backbone' controlling the national geo-body (cf. Thongchai, 1994), but all official information concerned with the functioning of disparate regions would pass through the city. This imagined centrality in a wider national order was expressed in urban design following the Prime Minister's directive that 'all other activities in the city including commercial, radiate from the Government Administrative Centre' (Department of Town and Country Planning, 1996: 1). A Government Precinct formed the northern section of the Core Area in the Putrajaya Master Plan. Running through the southern section of the Core Area was the 4.2 km Putrajaya Boulevard connecting the core island with the Government Precinct by means of the 'grand' Putra Bridge (see Figure 5.1, inset map). This led in turn to *Dataran Putra*, a ceremonial square for national celebrations. The Prime Minister's Office was to be located at the highest point along this central axis, overlooking *Dataran Putra*. With the other government offices 'arranged in a radial manner along the hill slopes' (*ibid*. 1) in the Government Precinct, the planning of the new administrative capital symbolised the centrality of state power within the national territory.

It was specifically post-colonial, independent state power that was performed here. On 31 August 1997, *The Sunday Star* included a special four-page advertisement to celebrate the fortieth anniversary of Malaysian independence (Putrajaya Holdings, 1997b) (see Figure 5.2). There is a point to be made here, first of all, about nomenclature: the first Prime Minister, Tunku Abdul Rahman Putra Al-Haj, remains a multicultural political icon and so the association of the birth of Putrajaya with independence suggested the benefits of the city for the whole nation. However, association with independence also served to attribute past and present urban problems to a legacy of colonialism. It was Malaysian dynamism and progress which, after forty years of political independence, had finally enabled architectural and technological liberation.[36] As with the Petronas Towers, there was a sense of the 'psychological liberation' sought in Vision 2020, expressed through the attainment of 'world class' (Putrajaya Holdings, 1997b: 24). But unlike the Petronas Towers – and partly as a response to criticism of this foreign-designed and built 'national' landmark – Putrajaya was said to be 'an embodiment of all that is Malaysian' and a 'showcase for local talent' (Tan Sri Azizan Zainul Abidin, President of Putrajaya Corporation, cited in *New Sunday Times*, 1999: 15).

The national credentials of the MSC cities were established in part by the companies leading their design and development. The concept for Putrajaya was formulated by the federal Department of Town and Country Planning with six Malaysian companies: Akitek Jururancang (Malaysia) Sdn Bhd, BEP Akitek Sdn Bhd, Hijjas Kasturi Associates Sdn Bhd, Perunding Alam Bina Sdn Bhd and Rekarancang Sdn Bhd (J. Lee, 1995). Putrajaya Holdings Sdn Bhd[37] was the company entrusted to develop the new administrative centre and did so in partnership with 'five of the nation's leading developers'

BIRTH OF A NATION
1957

40 years ago, our nation achieved her ultimate victory - breaking free from the bonds of her colonial masters, she emerged triumphant and proud as a sovereign nation.

BIRTH OF A CITY
1997

40 years later to the day, another momentous event is recorded in the history books of our nation. The birth of a world class, model garden city with intelligent features - PUTRAJAYA. An embodiment of all that is Malaysian, it stands as a glowing tribute to our achievements as a dynamic and progressive nation.

Figure 5.2 Putrajaya: Birth of a City; Birth of a Nation.

Source: Reproduced with permission from Putrajaya Holdings Sdn Bhd.

(Putrajaya Holdings, 1997b: 23).[38] Cyberview Sdn Bhd, the consortium which both owned and led the development of Cyberjaya's 2,855 hectare Flagship Development Zone was similarly comprised of some of Malaysia's biggest corporations.[39] There certainly was foreign participation – from NTT's involvement in Cyberjaya and the wiring of the corridor more generally to the subcontracting of aspects of the design process, such as landscape architecture in Putrajaya,[40] to non-Malaysian professionals and companies. Equally clear, however, was the downplaying of this participation in what were represented as projects of/for nation building. In the case of Putrajaya, in particular, proponents spoke of a conscious effort to limit foreign participation to specific areas where technology transfer was needed (John, 2000) in what was primarily 'a job for the locals' (Ng, 1997: 1). Widespread coverage in the national press and on television allowed national audiences to participate in the imaginative and 'real' building of the nation.[41]

This nation was also imagined in terms of a particular 'look' – a supposedly Malaysian urban aesthetic. One familiar mark of national landscape development in the MSC cities was the employment of a supposedly contextual or regionalist aesthetic. Allusions to 'local motifs' (Putrajaya Corporation, 1997) and a 'widely applied Moorish architecture' (Putrajaya Holdings, 1997a: 10) in Putrajaya evoked Malay-centred conceptions of national architectural identity considered in the previous chapter. The point was well exemplified by Tan Sri Datuk Seri Azizan Zainul Abidin: 'Architecturally, Putrajaya will be an indigenous city with a modern look ... it's a city of the future, but the local imprint is going to be very strong' (cited in Shareem, 1997c: 16). According to Marc Boey, Putrajaya, Cyberjaya and the MSC more generally were 'architects of nationalism', not only 'Malaysianising' a colonial landscape past, but also symbolically unifying fractured geographies of uneven global integration (Boey, 2002).

In a similar way, Cyberjaya, in particular, was bound up with attempts to define and realise an 'authentic' Malaysian 'natural' landscape. The 'Semi-nature Zone' of the city's 180 hectare Cyber Park, for example, was planned to include a special feature called the 'beyond garden'. This was an area in which 'the question "what is a tropical garden in the 21st century" will be studied, explored and experimented with' (Department of Town and Country Planning, 1997: 46). The expressed intention was to create a 'living laboratory' for the development of approaches to park and garden design in tropical Malaysia/Southeast Asia. This was a challenge which had also been taken up by the National Landscape Department (NLD). Established in 1996, and apparently reflecting the personal interests of Prime Minister Mahathir,[42] the NLD's role had come to include the formulation of national landscape guidelines (National Landscape Department, 1996). For the NLD, national landscaping proceeded by a simultaneous delving into past and future: peeling off layers of alien environmental transformation which mask the 'real' Malaysian identity; while, at the same time, carrying out research into ways of improving and speeding up the growth of 'indigenous' plant species. According to the Deputy-Director of NLD, Ismail Ngah, the Putrajaya and Cyberjaya projects were highly significant in helping to develop a national landscape policy.[43]

In Cyberjaya, in addition to on-going research in the 'beyond garden', there were attempts to restore and regenerate what was referred to as the 'original forest landscape' (Department of Town and Country Planning, 1997: 50). The rubber tree and oil palm plantation vegetation which covered much of the land on which Cyberjaya (and, for that matter, Putrajaya) was being built was described as, 'a holdover from the colonial period' (*ibid.* 50). It was to be replaced initially by orchards to replenish the soil and subsequently (in twenty to thirty years time) with the 'original' species of the area. This process was conceived of as something of a liberation from colonising plant species that had obscured authentic landscapes by themselves appearing to be 'natural' features of the Malaysian environment. The construction of Cyberjaya may thus be said to have imagined an inversion in which the imperial mentality of development became obsolescent Western Other to an advancing Malaysian modernity. The forest which, in colonial times, was cast as the negative opposition to plantations' modernity, an 'oasis of scientific order' (see Sioh, 1998: 147), would ultimately become a vital part of a post-colonial Malaysian city at the 'leading edge' of technological change. What is more, the involvement of NLD meant that these post-colonial national landscape innovations would be extended across the national territory.[44]

State representations of MSC also presented the possibility of extending technological innovations from the 50 by 15 km test-bed to all Malaysia(ns). Such technological utopian imaginings drew upon broader notions of the 'death of distance' (see Cairncross, 1997) effected by the development of information and communications technologies. Thus, in MSC discourse, it was specifically *national* life that would be freed from the constraints of space and the frictional effects of distance. Smart schools, for example, would benefit citizens in all areas because 'information can be disseminated throughout the nation' (Mahathir, cited in Krishnamoorthy and Surin 1997: 1). Electronic government was described in terms of the irrelevance of physical distance to interaction between state bureaucracy and citizens made possible by improved information flows and processes (Ibrahim and Goh, 1998). Government processes being 're-engineered' in Putrajaya would ultimately be 'rolled out' to all national citizens in all parts of the country, such that 'it will be possible to go to a kiosk in a shopping mall or use the PC at home to renew licenses and pay electricity bills in one simple session' (Mohamed Arif, 1996: 1). According to Prime Minister Mahathir, MSC was 'a pilot project for harmonising our entire country with the global forces shaping the Information Age' (Mahathir, 1998a: 30). 'In the end', the editor of *Investors Digest* projected, 'the entire country will be turned into a super corridor' (Harun, 1997: 1).[45]

The role of ICTs in integrating national space and promoting 'areal uniformity' (see Graham, 1998: 168) was performed in a so-called Teleconferencing Dialogue. Held in April 1997, this high-profile event saw the Prime Minister linked to 13,000 Malaysians in 28 locations across the national territory (Figure 5.3). Not only was each state represented, but 'those who posed questions came from all walks of life' (*The Sun*, 1997: 1).

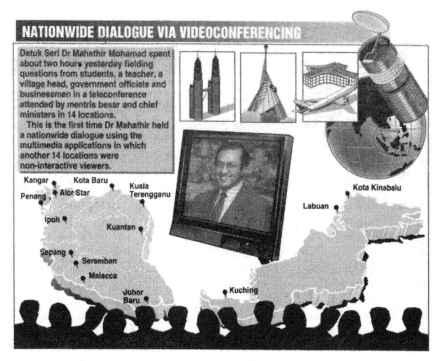

Figure 5.3 Nationwide dialogue via teleconferencing.

Source: Reproduced with permission from *The Star*, Malaysia.

Overcoming spatial difference within the national territory through new technology, therefore, was also imagined to be bound up with democratic and egalitarian possibilities (Asohan, 1997).[46] Such imaginings fostered a conception of national space as a discrete, undifferentiated unit that would benefit uniformly from technological advances pioneered in the MSC test-bed. In the Teleconferencing Dialogue, then, the information infrastructure enabling Malaysia(ns) to access multimedia networks was put to the service of a state imagining of an imminent intelligent national territory.[47] Supposedly revolutionary global technologies thus imaginatively reaffirmed and strengthened the existing (national) order.

The Teleconferencing Dialogue was in fact just one of a series of high-profile media events that (re)constructed the national centrality of the MSC. The development of the MSC cities, in particular, was marked by seemingly endless launches, opening ceremonies, symbolic dates and deadlines. Each step along the route to a national 'multimedia utopia' was broadcast as a spectacle with the Prime Minister almost always appearing as the charismatic focus: Mahathir not only uniting Malaysians through the wonders of cyberspace; but also now placing his palm on a monitor to activate a 'moonraker-type' ground-breaking machine; or planting a supposedly symbolic *merawan siput jantan* tree (*New Sunday Times*, 1997b). A more conventional advertisement from Putrajaya Holdings simply thanked their 'visionary' national leader for officiating the launch of the city (Putrajaya Holdings, 1997d: 5). Yet mediated MSC-scapes, I argue, themselves re-visioned the nation in ways that legitimised broader political transformation. If the high-tech test-bed ostensibly necessitated moves away from the defence of Malay economic interests, MSC discourses mobilised a vocabulary of 'bridge building', 'sharing', 'linkage', 'partnership' and 'connectivity' to (re)present a world in which success was dependent upon adopting and forging unity at various levels and where 'mutual benefits' could be derived from collaboration.

Media and other modes of display were thus technologies through which individuals and the population at large were able to envision new forms of citizenship. The fortieth anniversary of independence day advertisement on Putrajaya was part of a wider series of representations of the new city depicting an ideal national lifestyle, an aspirational model involving certain values and attitudes – to technology, to the family and social roles and to the environment – by which individuals could evaluate their own lives and conduct. To the extent that such conceptions of social health translated into the values and judgements of citizens, media representation induced citizen-subjects to realise themselves in 'intelligent' ways. In the technology fair or exhibition, so popular in 1990s Malaysia, visual subjectification was interwoven with other practices or techniques for the realisation of new ways of seeing and being. *Quality Urban Life '97 Exhibition and Multimedia Asia 1997*[48] was one of a series of exhibitions where an 'interactive' electronic Putrajaya stand allowed a multimedia experience of intelligent landscapes: new technologies of nationhood for intelligent citizens.

In the spectacle of these exhibitions, everyone could see themselves in the MSC cities of intelligent national futures. At the ground floor entrance to the two-storey Putrajaya stand, visitors filled in a registration form – name, address, ethnicity/nationality and monthly income – in return for which we were issued with a Putrajaya 'Smart Card'. The wall facing entrants to the first room was a screen on which was shown alternating English and Bahasa Malaysia versions of a video introducing the city:

> From Mother Nature a whole new city is taking shape, a city conceived and designed to be a total and complete city, one which will provide its residents and visitors with all the amenities and facilities so that they can live, work and play in an environment of a most pleasant, modern and progressive city. This is the city of Putrajaya: a city in a garden, an intelligent city of the next Millennium ... and one that is founded on Malaysian culture.[49]

A second room containing technical representations of the new city – the Master Plan and scale models – led up to the first floor. Upstairs was encircled by computer terminals which allowed the visitor to experience the intelligent city by means of their Smart Card. My exploration of the city began with an oblique aerial view. A large arrow pointed down to one of the low-resolution blocks which comprised a residential area: 'this could be Tim Bunnell's house'. Clearly this was, above all, a multimedia sales pitch – Putrajaya was, as the official marketing brochure put it, 'An Intelligent Investment' (Putrajaya Holdings, 1997a). Yet there is also a point to be made here about the way in which visitors' experiences of these electronic landscapes added coherence and legibility to the(ir) as yet un-materialised intelligent city. Subsequently, the Putrajaya web site (www.pjholds.com.my/) allowed the virtual visitor (and/or future resident) to enter and thereby visualise the street level at various sites/sights around the city by the click of a (left mouse) button. As Stephen Graham has suggested, new information technologies such as the Internet:

> ... actually resonate with, and are bound up in, the active construction of space and place, rather than making it somehow redundant. ... Material space and electronic space are increasingly being produced together.
>
> (Graham, 1998: 174)

Electronic as well as material landscape technologies performed something in new ways of seeing and being Malaysian.

In this chapter, I have considered the MSC as a space of government for the realisation of new forms of citizenship for high-tech times. Investment in physical and 'soft' infrastructure were (bio-)political strategies for the construction of an 'intelligent' Malaysia(n). The MSC cities Putrajaya and Cyberjaya were shown respectively to have been oriented to fostering an 'electronic' public sector and a technologically-innovative corporate knowledge worker. Yet I have also argued that in mid-1990s Malaysia, 'intelligent' came to extend beyond the provision or use of information and/ or multimedia technology. Rather, the MSC cities were bound up with broader conceptions of appropriate ways of seeing and being for success in an information society and economy. Appropriate 'national' constructions of technology were forged in opposition to the perceived social problems of Silicon spaces elsewhere as well as to existing 'social ills' in Malaysia. Focusing on the planning and initial construction stages of the MSC, I have considered underlying assumptions of a range of authorities, their governmental objectives and means to achieving them in and through the MSC cities.

The status of the MSC as a space of government, differentiated from modes of bio-political investment elsewhere in the national territory, fomented spatial tensions. We have seen in a previous chapter how official descriptions of MSC as a 'test-bed' were a means of limiting the political 'threat' of post-NEP economic liberalisation to one 50 by 15 km zone. In this chapter, I have considered the discursive elaboration of a rather different motivation for containment: MSC as a testing ground for potentially harmful 'global' socio-cultural as well as technological effects. The boosterist imaging of the corridor, domestically as well as for international consumption, meant that people and places outside MSC were perhaps more likely to be concerned at missing out on special treatment than thankful for being spared possibly deleterious consequences. We have seen how MSC was thus mediated as a national landscape with events such as the Teleconferencing Dialogue demonstrating possibilities for national-scale electronic integration and benefits. Nonetheless, the environments imagined as conducive for intelligent working and living remained sited within MSC. It is to emergent social and spatial divisions to which I turn in Chapter 6.

6 Beneath the intelligent cities
Socio-spatial dividing practices

Participants in the Teleconferencing Dialogue from across Malaysia questioned what, if anything, they had to gain from high-tech development concentrated in one 50 by 15 km corridor (Krishnamoorthy and Surin, 1997). This signalled popular awareness of the continued significance of spatial location despite authoritative imaginings of an 'end of geography' made possible by advances in information and communications technologies.[1] Certainly, the concentration of federal government high-tech investment in the MSC growth pole ran against regional development trends ostensibly oriented to a more 'balanced' national distribution (see Ghani, 2000). However, like other geographies of inclusion/exclusion, new divisions were a matter of socio-economic as well as spatial positioning. Some individuals and groups were considered worthy of incorporation into high-tech planning, even actively wooed from overseas as 'knowledge workers'. Other social groups were marginalised, even displaced to make way for the production of intelligent spaces. This chapter focuses on those people and places imagined as *not* belonging in would-be 'intelligent' Malaysian futures. It thus identifies socio-spatial dividing practices associated with rationalities of development in 1990s Malaysia.

In part, geographies of exclusion associated with the material construction of MSC were an extension of existing spatial development practices in Malaysia and elsewhere. Land required for infrastructure projects undertaken by state–private sector partnerships of various forms was acquired by the government and people displaced in the name of national progress. The cheapest land identified and targeted for (re)development was frequently that over which residents had limited land rights and/or legal claims to compensation. As such, the greatest social costs of transformation were borne by already socio-spatially marginal individuals and groups. New private sector-produced urban spaces were oriented to – and intended literally to accommodate – numerically and spatially expanding middle classes. The in situ residents – here I consider, in particular, plantation workers and indigenous Orang Asli groups – were financially excluded from privatised urban developments.

However, I suggest that there is also a specifically 'intelligent' dimension to the development of MSC and its intertwined social and spatial exclusions. In the first place, as we have considered in previous chapters, in the context of an imminent 'Information Age', large-scale, high-tech development was imagined as a necessity for national survival and success (see also Bunnell, 2002a). In other words, while Putrajaya and Cyberjaya shared important commonalities with previous large-scale infrastructure projects, MSC development was accorded additional high-tech urgency. The discourse of intelligent development for an emergent information economy and society thus legitimised both massive investment and associated social costs. In addition, however, exclusion from such projects was bound up with what might be understood as a 'moral geography' (see Matless, 1994) of intelligence. We considered in previous chapters how 'intelligent' came to be understood in normative rather than merely technological terms. Simply put, if the MSC cities were spaces for intelligent citizens, then those deemed unable to realise themselves in intelligent ways were implicitly (sometimes explicitly) rendered 'out of place' (see Cresswell, 1996). An intertwining of moral and geographical judgements was evident in the sites/sights of intelligent development. Pre-'intelligent' territories and their inhabitants had no place in authoritative imaginings of multimedia utopian futures.

This chapter, like the previous two, is divided into four main sections. I begin by considering critical perspectives on the practices and outcomes of new urban development highlighted in a conference held alongside the *Quality Urban Living Exhibition*. One such strand of critique concerned the exclusion of the majority of Malaysians from both decision-making processes governing social and spatial transformation as well as from emergent high-tech urban spaces. The remaining three sections then detail socio-spatial implications for marginal(ised) people/places in the MSC. This begins, in the second section, at Perang Besar plantation estate where Tamil workers who first came to Malaysia to work in British colonial rubber estates were displaced to make way for Putrajaya and excluded from residential development in the city. While this was partly a matter of economic calculation – the privatisation of premium urban(e) space – I demonstrate how a powerful moral geography also constructed plantation workers as unsuitable for intelligent national futures.

The third and fourth sections move southwards along Route 2020 to two sites below MSC's intelligent cities where transport infrastructure was under construction on land occupied by indigenous peoples. The first site is Kampung Bukit Tampoi where land was appropriated by the federal government for the construction of a new highway link. Like other Orang Asli groups in Peninsula Malaysia, the Temuan community[2] at Bukit Tampoi were cheap targets for transport 'development' in part because of a legacy of colonial land laws. I show how the developmental implications of the (post-)colonial nexus of race, space and the law were heightened with demand for land stimulated by the 'global' orientation and expansion of

Malaysia's primary city-region. The extension of transport connectivity, both within this urban region and externally, has been a clear planning priority in the globalisation of KLMA. Bukit Tampoi exemplifies more widespread associated eviction processes, the social costs of which have been borne disproportionately by Orang Asli and other politically and socio-economically marginal groups. The final site considered in this chapter also concerns displacement for global transport infrastructure. Kuala Lumpur International Airport (KLIA), as the southern 'node' of MSC, formed the end of the imaginative transect of Route 2020. KLIA also spelled the end of Kampung Busut, another Temuan village. I am concerned with rationalities of development which legitimised their dislocation. Official narratives of KLIA development traditionalised the Orang Asli in opposition to a planned, modern 'take off' of the(ir) land. Rather than incorporating Orang Asli into intelligent MSC spaces of Malaysian modernity, the swamp land resettlement site at Bukit Cheeding involved further spatial and social marginalisation to an 'aboriginal periphery' (see Ong, 1999: 218).

Urban utopia?

Running concurrently with *Quality Urban Life '97* was 'The South-South Mayor's Conference: Developing Solutions for Cities of the 21st Century', which marked the twenty-fifth anniversary of Kuala Lumpur City Hall. The official aim of the conference was to 'set the pace for hundreds of professionals, planners, engineers, architects and government authorities to have an interchange of invaluable ideas that will hopefully contribute to the betterment of urban life in the developing world' (Asian Strategy and Leadership Institute, 1997: np). Perhaps unsurprisingly given city authorities' boosterist image-consciousness (see Chapter 4), presentations on Malaysian urban development focused rather more on Malaysia's pace setting than on 'the woes that need to be addressed in the pursuit of urbanisation' highlighted in the organiser's introduction (*ibid.*). The conference did, however, provide a forum for critical perspectives on the nature of social and spatial development more broadly. These may be understood in terms of two key categories, both of which were readily applicable to Malaysia: the first concerns issues of privatisation and the exclusivity of private sector-led development; and a second category concerns the lack of public participation in the development process connecting to broader democratic demands articulated in Malaysia (as we have also seen in Chapter 4) in terms of 'accountability'. What linked these issues was a conception of *exclusion*: on the one hand, exclusion of everyday Malaysians from costly visions of modernity; and, on the other, exclusion from decision-making processes which shaped the(ir) built and 'natural' environment.

A four page article in the English language daily newspaper, *The Star*, entitled 'Urban Utopia?' mobilised conference themes for a powerful critical reflection on Malaysia's own way forward. Two questions formed the sub-heading for the article: 'Will quality urban living be privatised to the highest bidder? Will the new dream cities of Malaysia become the domain of the privileged?' (Sia, 1997a: 1). The illustration on the first page of the article contrasted grim, grey 'Gotham' city-style high-rise blocks and pollution-bellowing chimneys under clouds of smog or haze with a sealed and self-contained lifestyle utopia (Figure 6.1).[3] Thus, in part, critique of the effects of urbanisation and industrialisation on the natural environment was deployed to denaturalise the existing order of socio-economic and political development.[4] Entry to 'Luxuriaville', its clear blue skies, green spaces, detached houses and clean air was shown to come at a price – the

Figure 6.1 Urban utopia?

Source: Reproduced with permission from *The Star*, Malaysia.

'one million Ringgit' cost of entry was clearly intended to signify a sum well beyond the financial means of most Malaysians. While Putrajaya and Cyberjaya were lauded by their proponents as constitutive of a Malaysian future into which the whole nation could eventually be mapped, therefore, utopian urban developments were here re-envisioned as a privatised space excluding the majority of the population.

The conception of planned visions of urban development working to the exclusion of the majority of the population, of course, is hardly novel or unique to late twentieth century Malaysia. However, aside from undesirable experiences in 'the North' and elsewhere,[5] Urban Utopia was inspired, in part, by one of the projects on show at the exhibition – 'Sumurcity', the 'world's first covered city'. The proposal was not for a physically covered city – 'Sumurcity is actually only covered from a pedestrian's perspective ... there is no bubble covering the city but all the walkways will be covered, making it a pedestrian-friendly environment' (Klages, cited in Ngiam, 1997: 27) – yet the will to securitisation and exclusive segregation here was clear. As the master plan for Sumurcity was unveiled, Kuala Lumpur City Hall was reportedly undertaking a feasibility study for major streets to have air-conditioned walk-ways so that pedestrians paying 30 sen could walk in comfort (Kaur, 1997: 1). The implication that those unable or unwilling to pay would be left to other (not so pedestrian-friendly) environments evoked trends towards the privatisation of city streets noted elsewhere (see Jackson, 1998). Building upon such (Northern) trends, the very visual encoding of notions of 'futuristic', 'one-stop' self-containment in the Sumurcity logo may have been (mis)read as signs of escapism of the affluent few.

Marketing literature at the exhibition proudly announced that 'Sumurcity is masterplanned by renowned Klages Carter Vail & Partners from the United States of America'.[6] The international or, more specifically, American orientation of such projects is also encapsulated in Figure 6.1. Not only nomenclature – 'Luxuriaville' – but also the star-adorned sign-board is suggestive of US-style commercial development. It is perhaps significant that there appears to be little about Luxuriaville – or, for that matter, about the landscapes outside – which is distinctively 'Malaysian'. While ordinary Malaysians choked in their internationalist high-rise blocks and factories, the elite enjoy equally placeless lives in regularly-spaced housing architecturally reminiscent of North American suburbia. 'Country living' in fact became a focus of real estate development in Malaysia in the 1990s. Perhaps significantly, this did not refer to *balik kampung* (a 'return to the village' which many Malaysians, and especially Malays, make during public holidays), but rather to the construction of new 'resorts' on the outskirts of the city (*Real Estate Review*, 1993). Fears of loss of identity applied as much to this cosmopolitan real estate trend as to the modernist high-rise skyline of Kuala Lumpur (considered in Chapter 4). The new middle classes have been depicted as occupying a somewhat contradictory position in relation to such cultural politics: emerging as Western-educated producers and consumers of cosmopolitan cultural industries; but also reappearing as 'traditionalist' critics of Western-style development (Kahn, 1992). One contributor to a critical Muslim web site expressed concern that Malaysia's youth in particular was, 'hopelessly mesmerised by the charm of global consumerism as well as woefully inadequate to put up any resistance to it whatsoever' (Farish, 1996: np). In this light, the Sumurcity dome perhaps only fuelled growing fears about the cultural incarceration of next generation Malaysians.

The inclusion of a 'Hollywood Store' in the Sumurcity marketing logo was particularly apt as some critics suggested that MSC represented something of an open door to American cultural industries and values.[7] One Internet article on Mahathir's promotion of MSC to the American entertainment industry in Beverley Hills warned of 'the adverse effect that Hollywood will have on the morals of the youth should such an industry be allowed to flourish in Malaysia' (*Muslimedia Webzine*, 1997). More than this, the article suggested that the MSC was an invitation by the Malaysian government to allow Jewish film makers 'to realise their dream to destroy the Muslim world' (*ibid.*). The point here is not to reduce critique of the MSC to such shrill conspiracy theory. Rather, I suggest that the Muslimedia website exemplified a growing range of critical popular evaluations – including but certainly not only 'Islamic' – of hegemonic development objectives in Malaysia. The question mark in 'Urban Utopia?' thus denoted a diversity of socio-cultural and political problematisation.

Certainly, other critics considered that the main beneficiaries of plans for large-scale urban development were less likely to be Hollywood 'Zionists' than a small number of individuals whose companies had close connections with the Malaysian political elite. The issue here was not so much the desirability of privatisation *per se*, but rather the specific form it had taken in Malaysia. Gomez and Jomo (1997), for example, showed that the privatisation policy in Malaysia had allowed senior members of the ruling coalition to allocate lucrative national contracts to a politically well-connected corporate elite. These authors traced political economic networks linking large companies with senior UMNO figures such as Mahathir and Daim Zainuddin.[8] Unsurprisingly, this included companies involved in MSC urban development. Although planning authorities contended that Cyberjaya and Putrajaya were projects in which 'favoured' companies would *not* reap vast profits,[9] critical political economy had long diagnosed an 'unhealthy marriage' between politics and business as a feature of the Mahathir era (Gomez, 1994). While this marriage was legitimised, as we have seen (in Chapter 3), in terms of an ethnic politics of redistribution, the Malaysian political economy was one in which certain (mostly) Malay individuals had been enriched 'at the expense of the community as a whole, and certainly its poorer members' (Jomo, 1995: 6).

Perhaps behind the tinted glass of the vulgar yellow limousine entering Luxuriaville (in Figure 6.1), then, was one of the new Malay (they were not, in fact, all Malay) corporate stars and/or his (they were overwhelmingly male) political patron. For some, suggestions of self-serving elitism here were confirmed by the actions of the Sarawak Chief Minister, Tan Sri Taib Mahmud. The Chief Minister and his family were reported to have left the East Malaysian state for Europe on the day when a haze emergency was announced, a move which DAP leader, Lim Kit Siang, described as evidence of Taib's 'contempt for the people' (Lim, 1997b). The UMNO state leader enjoyed, 'clean and healthy air while leaving the people of Sarawak to fend for themselves while the state was choked with the worst air pollution in history' (*ibid.*).

There is a connection here with broader issues of governance. Not only were political elites able to escape dystopian Gotham but it was also they – as opposed to 'the people' – who controlled its development. In 'Urban Utopia?', Sia argued that most Malaysians were entirely excluded from urban decision-making processes: 'No matter how many times we conjure up dream cities to run away from congestion and pollution, it stands to reason that the only lasting guarantee of genuine quality urban life is when residents/owners have the power to influence planning decisions' (Sia, 1997a: 4). The article drew upon papers at the *South-South Mayors Conference* which invoked the Habitat Agenda.[10] Formulated at the Second United Nations Conference on Human Settlements ('Habitat II') in Istanbul, the Habitat Agenda emphasised the importance of the role of non-state actors in the shaping of the urban environment (United Nations Human Settlements Programme, 1996). Thus, to an extent, the Malaysian auto-critique in 'Urban Utopia?' was bound up with (and legitimised by) supranational discourses and vocabularies of 'good governance'. The assertion of individuals' and communities' rights to shape their own urban future, for example, was couched in terms of citizens as 'shareholders in development' (Nicholas You, co-ordinator of the UN Centre for Human Settlement, cited in Sia, 1997a: 4).

However, critique also took the form of a specifically Malaysian discourse of political accountability already alluded to in previous chapters: 'the general notion of accountability ... perhaps more accurately reflects debates on the limits and expectations of power than other terms we might use that are rooted solely in Western institutions and in the ambitions of those who champion them' (Harper, 1996: 251).[11] According to Tim Harper, the concept extended from: recent accounts of nationalist struggle in pre-independence Malaya;[12] to the formation of citizens' organisations in opposition to Mahathir's authoritarianism in the 1980s; to outrage at the corrupt, immoral or decadent lifestyles of contemporary political rulers. The argument in 'Urban Utopia?' that Malaysian local council elections should be reinstated[13] was also couched in terms of accountability. One retired academic authority is reported as suggesting that 'appointed local councillors are not accountable to the public and some are arrogant and behave like autocrats' (Dr Cheah Boon Kheng, cited in Sia, 1997a: 4).

A discourse of accountability thus imagined new forms of governance in Malaysia which would give individuals and residents groups outside Luxuriaville greater control over the(ir) urban environment.

'Urban Utopia', in sum, was a powerful critical reflection on the official planning and marketing of the quality of Malaysian urban living. On the one hand, there was something hopeful in the Habitat II-inspired subheading on the last page of the article: 'Together we can make a difference' (Sia, 1997a: 4). The very existence of the article in a mainstream newspaper suggested space for mobilising citizens towards progressive urban futures. On the other hand, the Malaysian present exhibited few signs for alternative ways forward. As Sia suggested, 'Transparency, accountability, citizens' participation and residents' rights – none of these figured in the Quality Urban Life Exhibition which, by and large, presented a future based on developers' showcases and public relations glitz' (Sia, 1997a: 4). Private sector imaginings obscured but surely promised only greater socio-spatial division. In the words of another speaker at the South-South Mayors Conference, Dr Lim Teck Ghee, a United Nations Regional Adviser on poverty alleviation: 'The private sector may build high-rise luxury blocks and marinas but not affordable housing' (cited in Sia, 1997a: 2).[14] Certainly, the Malaysians who stood to benefit from an MSC urban utopia were not the Federal Territory's squatters or inhabitants of the city's low-cost, high-rise public flats. Nor were they the plantation workers living on sites earmarked for MSC's intelligent cities. It is to the displacement of plantation workers for the construction of Putrajaya which I now turn.

From Perang Besar to electronic administrative centre: moral geographies of Putrajaya

Up until the mid-1990s, the 4,581 hectare area covered in the Putrajaya Master Plan was dominated by oil palm and rubber plantations. Putrajaya was frequently referred to as a 'greenfield project' (see, for example, Mohamed Arif, 1996: 1). The promotion of MSC more generally as 'empty' space imagined the unhindered construction of a high-tech Malaysia in/for would-be investors. In planning terms, the southern growth corridor was commonly considered a 'natural extension of Klang Valley's urban fabric' (Singh, 1997b: 7) relieving increasingly 'overcrowded' urban conditions. As we saw in the previous chapter, however, MSC development was also understood in terms of a territorial 'Malaysianisation' through the replacement of colonising plant species. This process of landscape nationalisation can be traced back to the 1980s with the Malaysian government's purchase of British plantation companies. The most famous of these involved a 'dawn raid' on the London stock exchange for Guthrie Corporation in September 1981 (Khoo, 1995: 5).[15] The following year saw the take-over of Harrisons and Crosfield – subsequently renamed Golden

Hope Plantations Berhad – which owned the majority of the land eventually acquired for the construction of Putrajaya.[16] As industrialisation gained momentum in the Mahathir era, celebration of Malaysia having taken control of its 'own' resources was increasingly accompanied by conceptions of plantation agriculture as a sign of commodity-dependence. Goals of economic upgrading, therefore, became intertwined with an imperative of ecological decolonisation. As discussed in the previous chapter, MSC development was thus conceived of as replacing an obsolescent commodity-based economics of imperial modernity with indigenous high-tech innovation. Particular people as well as places were rendered obsolescent in this intelligent (re)imagining.

The new intelligent administrative centre was not constructed on empty plantation space. In addition to three villages on the site, the Environmental Impact Assessment for Putrajaya and surrounding areas – carried out by *Universiti Pertanian Malaysia* (UPM) – identified some 2,400 or so residents in four plantation estates (UPM, 1995).[17] It is specifically the experiences of these plantation estate residents which I consider here since they, unlike inhabitants of the local *kampungs*, experienced particular problems associated with their (lack of) land ownership. Given that plantation communities had no formal rights to the land in which they had invested as their homes – in some cases for four generations – they had much in common with urban squatters. Indeed, the two groups campaigned together at a Labour Day Rally in 1994 submitting a joint memorandum to the Prime Minister which claimed that 'both communities ... have been neglected in the sharing of the nation's wealth' (Chandran, 1994: 10). In many cases, the two communities were, in fact, one and the same 'forsaken lot' (*ibid.* 10) since a significant proportion of the squatter population in the state of Selangor, in particular, in the 1990s was comprised of ex-plantation workers (Sivarajan, 1995).

However, if the number of squatters in the Klang Valley proliferated as a result of rural–urban migration for post-independence (and especially post-1970) industrialisation, plantation workers' precarious socio-spatial position stems back to the colonial division of labour (see Chapter 3). Tamils from the lowest social classes in south India were brought into Malaya from the late nineteenth century to work on the British-owned rubber plantations. In the peak year, 1927, 120,000 arrived as cheap rubber estate labour (Brockway, 1979: 161). Smaller numbers of middle-class Tamils were also recruited to act as foremen in the estates as well as to work in the English-speaking civil service. While it is now widely recognised that Tamils were characterised as suitably 'docile' for estate labour in terms of colonial racial stereotypes (Ramasamy, 1994; Stenson, 1980), labour shortages which necessitated immigration were, in part, the result of a pre-existing spatial divide among racial/ethnic groups in Malaya. Malays had their own land and Chinese workers were employed in Chinese-owned plantations (Sioh, 1998). As a result, while Tamil migrant workers came to perform a distinct function in the colonial Malayan division of labour, they owned no land in racially-divided colonial space; their place in colonial society amounted to little more than their employment. To the extent that corporate-style agriculture introduced by the British continued to predominate after Malaysian independence, this uncertain position remained largely unchanged in late twentieth century (post-)colonial Malaysia.

'Indian' Malaysians – and Tamil plantation workers in particular – increasingly consider themselves to have been marginalised further in a politically and culturally Malay-dominated post-colonial state.[18] As a political minority, Malaysia's Indian population is said to have 'encountered problems in getting what Malaysians of other ethnic origins are reaping' (Ramasamy, 1997: 8).[19] During the NEP period, while the proportion of privately held national wealth in both Chinese and Malay hands increased – largely at the expense of foreign holdings – the Indian share actually fell from 1.1 per cent to 1 per cent (Jayasankaran, 1995b).[20] The situation is such that some politicians, including even the leader of the Malaysian Chinese Association (MCA), suggested that there may be a need for an 'NEP for the Indians' (Dr Ling Liong Sik, cited in Jayasankaran, 1995b: 26). At one level, the small size of the Indian population has served to make Malaysian Indian Congress (MIC) very much a junior partner in the ruling *Barisan Nasional* coalition and this weakness has been exacerbated by 'factional squabbles and financial scandals' (Ramasamy, 1997: 10). Yet working-class estate labourers, ostensibly represented by south Indian-controlled MIC, have been doubly marginalised through a party dominated by an urban and business professional class that has tended to focus on its own interests to the neglect of plantation issues (Ramachandran, 1994). Thus, even high-profile entrepreneurial Tamil success stories – most notably T. Ananda Krishnan of KLCC fame (as discussed in Chapter 4) – have only brought into sharp contrast 'a disadvantaged community that needs assistance' (Ramasamy, 1997: 10).

The following is an excerpt from a poem published in the monthly journal of the Consumers' Association of Penang, to 'celebrate' the 100th anniversary of the plantation industry in Malaysia:

> ... I have no home in my own country
> Estate sold at boss' own fancy
> Years we had slogged so tirelessly
> Now, shunned, shooed, so shamelessly.
>
> Sold – like land and rubber trees
> Eviction order given so efficiently
> Long-service forgotten conveniently
> Displaced ... family in great misery
>
> No house, no job ... uncertainty
> Never mind, there is Vision 2020

Squatting amidst vice in the city
A non-priority ... all eyes on the MSC ...
(*Utusan Konsumer*, 1997: 20)

A situation in which the workers' destinies were at the 'fancy' of plantation bosses was inherited from the colonial period. But a post-colonial context in which plantation workers have 'no home' in their 'own country' also suggested that their position had not improved with citizenship rights at independence. As we have seen, however, homelessness here is to be understood more literally too: given that the relationship between the workers and estate management in Malaysia has included housing provision, the (re)development of plantation community land resulted in their loss of shelter as well as loss of employment (Sivarajan, 1995). That the threat of change is denoted in the poem by government plans – Vision 2020 and MSC – implied overlap, or perhaps even complicity, between business and state in workers' displacement. In fact, through the shareholdings of Permodalan Nasional Berhad (PNB) and Amanah Saham Raya, the government was also effectively in control of the biggest plantation companies (see also Arutchelvan, 2000).

In the mid-1990s, almost two-thirds of the shares in Golden Hope Plantations Berhad were held by PNB, the government's trust agency for *bumiputeras* (Gomez and Jomo, 1997: 38). This intertwining of the state and big business may be said to have underscored plantation workers' political and economic powerlessness. Their century-long role in national development was overlooked in a scramble to maximise corporate and (ostensibly) Malay ethnic redistributional returns. Workers were sold, 'like land and rubber trees' (*Utusan Konsumer*, 1997) as part of an environmental take-over. The government took possession of Perang Besar Estate in December 1993 under Section 8 of the Land Acquisition Act in return for which Golden Hope reportedly received a 'windfall' of about RM 450 million in compensation for the sale of the estate (Tan, 1994).[21] Meanwhile, Golden Hope itself – as one of the 'national' companies involved in the Cyberview consortium – announced plans for the development of a Corporate Campus in neighbouring Cyberjaya (*The Star*, 1997a: 14). Another plantation company in which the government holds a sizeable share, Malaysian Plantations Berhad, emerged as one of the developers for Putrajaya. With 'all eyes on the MSC' the welfare of plantation workers was even less likely than ever to be seen as a national priority. The construction of intelligent MSC spaces in plantation land at Putrajaya resulted in estate workers' displacement off site as well as out of sight.

With 177 families, or around 1,500 people, and covering more than 1,700 hectares, Perang Besar was the largest of the Putrajaya estates (see Figure 6.2).[22] Operations began in the 1930s and, at the time when employment termination notices were given out in 1995, consisted of crops of oil palm and cocoa as well as rubber.[23] The law, as it stood, required

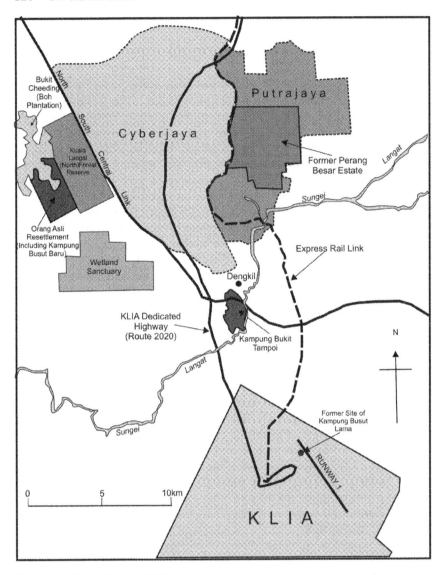

Figure 6.2 Along Route 2020.

the employer to pay only twenty days of wages per year of work to the displaced worker (Sivarajan, 1995: 52). However, in common with the other Putrajaya estates, there was considerable wrangling over estate residents' compensation involving not only the plantation company and the federal government,[24] but also the Selangor state government (because the relocation site was likely to be in Selangor Darul Ehsan), the Malaysian Indian Congress (MIC) (because the workers were classified as 'Indians'), the National Union of Plantation Workers (NUPW) and NGOs.[25] After Golden Hope ceased operations at Perang Besar, residents sought work outside the estate while negotiations over compensation and resettlement continued.

By 1997, conditions in the estate had deteriorated. Public transport had become less frequent and more expensive. A reduced bus service from the nearby town of Kajang (see Figure 6.2) plied a route across a newly-surfaced section of road. Prior to Putrajaya, when Perang Besar was still 'a place hardly known even to people in the Klang Valley' (Rajah and Perumal, 1994: 3), residents had campaigned for years to get the surface improved.[26] In addition, electricity – provided by Putrajaya Holdings after the estate company left with the original generator – was continually being cut off[27] and water pumped from an increasingly polluted river looked, but apparently did not taste, like tea.[28] Inflation associated with the 'booming' development in the area left Tamil plantation communities more generally wondering if they could afford to celebrate Deepavali (Sia, 1997b). Nonetheless, despite these conditions, what appeared to worry residents most was the prospect of relocation.

Resettlement into low-cost flats was imagined by residents as a threat to community health and well-being. When the local union official visited to ask residents about housing preferences, they unanimously requested low-rise housing (Sia, 1997b). One Perang Besar resident, Komathi, did not think that she was suited to high-rise life. She had a brother living in a high-rise block on Jalan Pantai in Kuala Lumpur. When there was a death in one of the families who live on the fourth floor, she said, Muslims on the floors below would not allow the family to take the body to the flat upstairs. In contrast, the issue of encroaching, she said, never arose in the estate where space was communal. In the case of funeral ceremonies, for example, tents would be erected along the frontage of a number of adjacent houses.[29] High-rise was thus imagined as synonymous with a privatising fragmentation of space. In contrast, the public communal space of the estate was considered particularly important for the welfare of the elderly and children. In the estate, according to Nithya, 'the old people have a community to look after them and they can walk around anywhere'. Similarly, especially for 'poor families', 'both parents have to work and can't afford to have people to look after the children'. This was partly an issue of familiarity 'in the village where all strangers are recognised immediately and everyone knows everyone else' in contrast to the imagined anonymity of the city. However, residents also drew upon broader discourses of urban social ills in Malaysia. Keeping an eye on the children was not just a matter of safety, it was also a means of enforcing traditional social codes which prevent children from getting involved in 'unhealthy' activities.[30] Prevention here was not simply an issue of surveillance. Rather, the estate was also understood as a community – focused, in particular, around the (Hindu) temple – in which everyone had a role, a place.[31] Communal plantation life, like utopian MSC urban development, was valorised in opposition to the perceived ills of Malaysian urban life.

Residents did not simply romanticise estate life. On the contrary, it was precisely because of the hardship suffered by old people in particular in the plantations that they should not be 'shunned, shooed' (*Utusan Konsumer*, 1997).[32] As Nithya pointed out, 'many of the people in the village have had very hard lives – they are not office workers and are not so strong in old age'. Hard work had brought progress symbolised by the 'third generation' houses in which they were living as work on Putrajaya began, but this was a trend which she perceived would be reversed if the community was forced to move to walk-up flats: 'they have laboured all their lives and now they are being asked to labour in their old age'. As part of the broader issue of entitlement, there is a strong sense of connection with the land here, of having earned the right to benefit from its 'national' redevelopment; 'their hard work in Putrajaya should be rewarded with houses in Putrajaya'.[33]

The Putrajaya Master Plan included some 10,000 low-cost housing units, although their location in the city and the form they would take was uncertain in 1997, other than that they would be 'world-class' and 'wired'.[34] The government had earlier made it mandatory for developers to build at least 30 per cent 'low-cost' units (that is, not exceeding RM 25,000) in all housing projects (Ghani and Lee, 1997). Certainly, residents of Perang Besar and the three other estates were not 'rewarded' with any such units in the new city. Plans to move the Perang Besar community to Kuala Selangor and subsequently to Rawang[35] were rejected in part because of their distance from Perang Besar and the resultant disruption in terms of schooling and employment. However, residents also expressed a desire to make the most of opportunities in the(ir) area in the future. A new site near Dengkil (see Figure 6.2), where five blocks of 80 units would house all four estate communities,[36] would be close to Putrajaya, but why was no provision made for them in the construction of the Intelligent city? It later emerged that the 10,000 units (or 15 per cent of the 67,000 total number in the city) would be 'affordable housing' at RM 49,000.[37] Low-cost housing, like former plantation workers, did not belong in the 'intelligent' federal government administrative centre.

Exclusion here extended beyond affordability. There were intertwined ethnic and class politics at work in processes of plantation worker eviction. The national level ethnic quota system favouring Malays in government jobs had powerful local implications with the development of Putrajaya. Since the new electronic government centre entailed the relocation of the federal administrative machinery, it was expected that 'affordable' housing would be allocated to personnel from the Malay-dominated civil service. The MIC reportedly told residents that the 'Malay government' did not want to give any compensation, thereby implying that only their political efforts had secured the flats in Dengkil.[38] Residents, however, appeared to be less likely to resort to ethnic 'explanations' than their erstwhile political representatives. Indeed, residents I spoke with suspected that this was only a ploy to get them to leave and, besides, the local MIC leader was away – reportedly at his home in Johore Baru.[39] Meanwhile an article reporting MIC leader, S. Samy Vellu's, speech at his party's general assembly in Kuala Lumpur urged Indians to 'face the reality of the Multimedia Super Corridor and its challenges by equipping themselves with the right education, attitude, intelligence, creativity and competency' (*New Sunday Times*, 1997c: 3).

The significance of Samy's speech was precisely the way in which it legitimised the exclusion of plantation workers from intelligent spaces. Samy constructed 'Malaysian Indians' and especially those 'trapped in the plantations' as not being 'ready for MSC', for not possessing the appropriate education or attitude for high-tech times (and places). For Samy, 'this sub-culture is a mind-set that is besieged by hopelessness and helplessness'. While acknowledging 'lack of opportunities to pursue education and employment', according to Samy, plantation workers, along with urban squatters, had 'alienated themselves from the mainstream Malaysian society' (cited in *New Sunday Times*, 1997c: 3). Attempts to understand Indian social problems, particularly in the media, frequently invoked notions of racial disposition or mentality. A stereotype of inherent Indian aggressiveness, instability and/or criminality was sadly ironic given that British colonial recruitment of Tamil labour was premised on an equally racist belief in their docility and submissiveness (Ramasamy, 1994). The wider regime of representation which cast sections of the Indian community as problematic, even if ostensibly mobilised to advocate greater political efforts to improve their social position – Samy spoke of policy action for a 'disadvantaged' and 'delinquent' community as well as of an MIC 'social recovery programme to address social ills' – rendered such people and places unsuitable for 'intelligent' development. Samy thus perpetuated a moral geography rationalising the displacement of 'Indian' plantation workers from obsolescent and degenerate spaces. Their land was infrastructurally regenerated to facilitate the lives of supposedly more intelligent others.

Urban connectivity and the globalisation of Orang Asli land

From Malaysia's intelligent cities, the proposed Route 2020 plied a course past Dengkil where the displaced plantation workers were to be rehoused, eventually intersecting with another new road connecting the North-South Highway to the airport under construction at Sepang (at the southern end of the MSC). This North-South Central Link in turn also served to integrate the new southern corridor with the existing Klang Valley urban region. Less than a kilometre east of the planned intersection with Route 2020, adjacent to the Langat River, the North-South Central Link passed through the northern section of Kampung Bukit Tampoi (see Figure 6.2). The houses of seven families of the Temuan community there were destroyed in March 1996 for the new link road. The globally-oriented expansion KLMA thus impacted upon another politically marginal group without formal land title. This section is concerned with the spatial processes and politics of 'global' development *vis-à-vis* the Orang Asli and the Temuan of Bukit Tampoi in particular.

Orang Asli or 'original people' is the official term for the diverse indigenous communities of Peninsula Malaysia. This diversity has been anthropologically and administratively reduced to eighteen official sub-groups which are in turn conveniently organised into three major classifications, Negrito, Senoi and Proto-Malay (Nicholas, 2002).[40] The community at Bukit Tampoi are Temuans who form part of the broader Proto-Malay classification. However, in terms of national development, state authorities do not appear to see Temuan as distinct from other *asli* groups.[41] A crucial dimension of commonality relates to a politics of indigeneity in the (post-)colonial nation-state. Three different sections of the Malaysian population are in fact recognised as 'indigenous': the politically (and, in the case of the Peninsula, numerically) dominant 'Malays'; and the 'natives' of Sabah and Sarawak, an umbrella term for various groups in the East Malaysian states on Borneo; as well as the minority first peoples of Peninsula Malaysia who, since 1960, have been termed 'Orang Asli' (see Nicholas, 2002). As we have seen (in Chapter 3), especially since the early 1970s, economically- as well as politically-privileged indigeneity has been denoted by the term *bumiputera* and (in Peninsula Malaysia at least) this is synonymous with 'being Malay'[42] – that is to say *Melayu* as opposed to *asli* (Zawawi, 1996a). Complex political slippages between *bumiputera*, *Melayu* and Orang Asli have been a double-edged sword for Peninsula Malaysia's first peoples. On the one hand, Orang Asli are theoretically able to claim the special rights which the federal Constitution provides specifically for Malays in Peninsula Malaysia. On the other hand, implicit and sometimes overt Islamicising attempts to subsume Orang Asli under 'Malay' have undermined rights to cultural self-identification.[43]

It was in part this problematic post-colonial positioning that stimulated efforts towards Orang Asli political organisation. *Persatuan Orang Asli Semenanjung Malaysia* (POASM, 'Peninsula Malaysia Association of Orang Asli') was established in 1977 by educated members of various Orang Asli groups spurred initially by proposals to rename the group '*Putra Asli*' ('aborignal Princes'). Two further spurs for the formation and expansion of POASM concern issues of political representation and supposedly pastoral protection. First, although Orang Asli do have one Senator, this is a post appointed by the federal government. Orang Asli individuals and groups have been a largely 'unrepresented minority' in the ethnic politics of post-colonial Malaysia (Karim, 1995: 20). Second, state responsibility for Orang Asli 'protection and advancement' (Ministry of the Interior, 1961: 3) is in the hands of a specific government agency, the *Jabatan*

Hal Ehwal Orang Asli (JHEOA, 'Department of Orang Asli Affairs'), originally established in the 1950s (as 'The Department of Aboriginal Affairs'). This federal government department – popularly referred to as simply 'JOA' (*Jabatan Orang Asli*, 'Orang Asli Department') – was identified by the POASM leader in 1997 as a key cause of Orang Asli problems. For Majid Suhut, in the light of state development priorities, the 'J' in JOA had become '*jual*' ('sell') – the selling (out) of Orang Asli interests and welfare.[44] Ironically, by providing diverse groups with a 'single entity on which to focus their grievances' (Nicholas, 2002: 125), the JOA has played a key role in the formation of 'Orang Asli' as a political identity expressed through POASM and other organisations.[45]

The perceived shortcomings of JOA are nowhere more clearly evident than in relation to issues of land. Like the plantation workers considered in the previous section, Orang Asli post-colonial placing is a legacy of legal systems established by the British colonial power (Williams-Hunt, 1995). The colonial Aboriginal People's Act (or 'Act 134') empowers authorities to gazette land exclusively or predominantly inhabited by Orang Asli as 'aboriginal area' or 'aboriginal reserve' allowing occupation without payment of land taxes (Malaysia, 1994). Yet, spurred historically by security concerns during the Communist 'Emergency' (see Chapter 3), the developmental emphasis in independent Malaysia has been on moving Orang Asli communities out of jungle areas. Under the direction of JOA, the ostensible aims have been: to resettle Orang Asli into villages provided with medical facilities; to provide them with access to mainstream educational institutions; and to encourage them to change their lifestyles in order for their 'integration' with Malays to be successful (Zahid, 1990). Apart from the cultural spectre of assimilation here (see Karim, 1995), Orang Asli resettlement in the 1990s appeared to be more a matter of making space for 'development' than a means of improving their social or economic welfare. Certainly, state governments (who control land matters) have made full (ab)use of their power under Act 134 to revoke wholly or in part any declaration of land gazetting. In the state of Selangor, as elsewhere in the 1990s, there was a marked reduction in gazetted Orang Asli reservation land (Nicholas, 2000). In the break-neck urban and economic growth of 1990s Malaysia, JOA facilitated the appropriation of Orang Asli land for 'national' development. This was particularly the case in prized territories around existing urban centres such as the expanding KLMA.

Orang Asli land, such as that at Bukit Tampoi, was a cheap target for (re)development. Orang Asli Reservation land – as distinct from Malay Reservation – does not include individual ownership titles. 'Just' compensation for land (in line with section 12 of Act 134) or fruit trees (section 11) is determined by the 'Commisioner of Orang Asli Affairs' (or what is today the Director-General of JOA) (Malaysia, 1994). The appropriation of Orang Asli land has thus typically meant compensation only for fruit trees and other crops, and certainly well below the market value of the land which would be paid if it was acquired from owners under the Land Acquisition Act. As such, it is unsurprising that Orang Asli land has been targeted for development in KLMA and elsewhere. The flexible cartographies of colonial capitalism have been fully exploited by the (post-)colonial developmental regime. At Bukit Tampoi, while forest and river resources had long been depleted by environmental change (Sham, 1989), an initial globalising encroachment involved the degazetting of an elevated plot of 1.21 hectares to make way for a radar site for the new international airport under construction further south.[46]

It was the subsequent construction of the North-South highway link to the airport that led to eviction of Temuan from 15.57 hectares of land in the settlement. This land was acquired by the Selangor state government in March 1996 on behalf of the federal government which, through the *Lembaga Lebuhraya Malaysia* ('Malaysian Highway Authority') had contracted the government-linked United Engineers (Malaysia) Bhd to build the new section of road. The authorities treated the Bukit Tampoi Temuan community as squatters or trespassers on state land. Their houses, two communal buildings as well as crops and fruit trees were destroyed in police operations on 22 and 27 March with support from the Federal Reserve Unit. Compensation was offered only for loss of crops, fruit trees and housing (that is, the building structure rather than the value of the lands lost) (Jerald Gomez and Associates, 2002). However, the road passed through land which was variously: 1. gazetted aboriginal reserve; 2. scheduled for gazetting; and 3. ungazetted but customarily occupied by the Temuan of Kampung Bukit Tampoi.[47] In October 1996, evicted community members began court action to vindicate their rights to place and obtain appropriate compensation. In the mean time, construction of the road continued in and through the settlement (see Figure 6.3).

Socio-spatial dividing practices associated with the construction of the North-South Central Link at Bukit Tampoi extend beyond the literal dissection of the settlement shown in Figure 6.3. While it is visible and audible from the *kampung*, there is no direct access to the new road at Bukit Tampoi. A trend identified in urban and regional studies more broadly is that of 'strictly hierarchical highway systems with highly limited access points' which are associated with increasing 'horizontal segregation of uses within the metropolitan region' (Graham and Marvin, 2001: 120). Like other major expressways, then, the North-South Central Link connects globally 'significant' sites, while bypassing adjacent localities. People and places are configured in highly differentiated ways in relation to logics of global connectivity. In greater Kuala Lumpur as elsewhere, global transport as well as information infrastructure serves to 'splinter' urban space in increasingly complex ways (*ibid.*). For Bukit Tampoi residents, the new link road constituted a divide or barrier rather than an avenue for global interconnection.

Figure 6.3 Global connectivity and new divisions.
Source: Author's photograph, 2003.

Yet, somewhat paradoxically, the legal case for compensation drew upon extra-local and even extra-national connections. While Orang Asli groups have been tied into extensive networks of trade for centuries before the current round of globalisation,[48] they have more recently been enrolled in transnational networks of 'indigenous peoples' (see also Smith *et al.*, 2000). In Malaysia, the Centre for Orang Asli Concerns (COAC), coordinated by Dr Colin Nicholas, has played a key role forging such connections. Nicholas notes how, in May 1992, indigenous peoples' movements and organisations in Asia established the Asia Indigenous Peoples' Pact at an assembly in Bangkok (Nicholas, 1996: 2). A key shared experience motivating this pact was a 'feeling of loss of control over their lands' (*ibid.* 2). This of course further connects Orang Asli to the experiences of indigenous peoples and places beyond 'Asia', as performed through the celebration of World Indigenous Peoples Day in Malaysia. Such 'global' experiences were drawn upon in the case made for Sagong Tasi and other displaced Temuan from Bukit Tampoi against the government. In particular, indigenous people's cases for 'native title' in other nation-states added legitimacy to Orang Asli land claims. Yet a perhaps more familiar claim for the plaintiffs was that the globally-oriented government had failed to perform its fiduciary obligations to the Temuan of Bukit Tampoi. JOA members were present at the evictions through which Orang Asli were disconnected from land occupied by them for generations (Jerald Gomez and Associates, 2002).

The end of the road: KLIA and Kampung Busut (Lama)

From the intersection with the North-South Central Link that divided Kampung Bukit Tampoi, Route 2020 was intended to extend a further 5 km into the Kuala Lumpur International Airport (KLIA) site. By 1997, this had become known as the southern 'node' of the MSC. K. L. International Airport Bhd's own earlier marketing literature preferred slightly different metaphors – referring to the new airport as 'the new transportation hub for the Asia-Pacific region' (K. L. International Airport Bhd, 1997) or 'the pivot of ASEAN' (K. L. International Airport Bhd, 1994a) – but the planning intent was clear: if new roads such as the North-South Central Link integrated key nodes in the extended metropolitan area, the new airport was intended to make KLMA central to a supranational region of interconnected cities. Yet, as for other development projects that sought to incorporate supposedly intelligent citizens into an emergent network society, KLIA served to peripheralise other Malaysians. In this final section, I consider the case of Kampung Busut, a Temuan village in the Sepang district affected by the RM 8–9 billion airport project at the southern end of Route 2020 (refer to Figure 6.2).

By assimilationist state systems of evaluation, Kampung Busut was a very successful community. The village had an UMNO branch (Nicholas, 1991) and JOA had even provided a *surau* for converts (Engineering and Environmental Consultants Sdn Bhd *et al.*, 1993: E3–33). According to the Environmental Impact Assessment (EIA) compiled in 1992–3, the community was 'experiencing the Malaysian mainstream social and economic life' while at the same time continuing to 'identify themselves as Orang Asli by practising their traditional social and family structure' (*ibid*. E3–33). Positive authoritative evaluation of Kampung Busut (and neighbouring Kampung Air Hitam) in the EIA was in part precisely because it 'does not reflect an aboriginal environment' (*ibid*. E3–35). The community at Kampung Busut was not living in or off the forest, but rather gained much of its income from the sale of oil palm, rubber and seasonally from durians. This economic activity, in Sepang as elsewhere, was in fact necessitated by the depletion of surrounding forest areas.[49] Nonetheless, residents were said to 'have successfully transformed their "nomadic life"' (*ibid*. D2–20).[50] Two years before the EIA, Kampung Busut had been nominated by the district political leaders as a 'model village' for tourism development (Zawawi, 1996a).

The model Orang Asli settlement at Kampung Busut formed part of the 8,400 hectares of alienated land to be acquired by the government for the new airport (Engineering and Environmental Consultants Sdn Bhd *et al.*, 1993: E3–20). Given geo-histories of territorial attachment dating back to the 1810s (Ramli, 1991), residents were very anxious about the prospect of resettlement. A Village Action Committee headed by the Chairman of the village UMNO branch was formed to discuss compensation and for dealing with the relevant authorities. The EIA recognised a list of crops compiled by the action committee which should be compensated, and identified a 700 hectare site at Bukit Damar (near Perang Besar Estate) as suitable resettlement land (Engineering and Environmental Consultants Sdn Bhd *et al.*, 1993: E3–36). Among the apparent merits of this new site were the fact that it was 'a fully developed agriculture area which can immediately sustain the economic needs of the community by way of matured oil palm and rubber' and its location near to 'growth areas which can be a good source of employment for the community' (*ibid.*, 1993: E3–37). According to a documentary produced by Kuala Lumpur International Airport Berhad, the Kampung Busut community had nothing to fear: Orang Asli were to be mapped into a broader 'lift-off to the future' effected by the new airport hub (K. L. International Airport Bhd, 1994a).[51]

The eventual relocation site for the two KLIA Temuan communities was characterised by neither pre-existing profitable agriculture nor easy access to opportunities created by the airport and other new developments.[52] Nor was the 400 hectares of land[53] which had previously formed part of the Kuala Langat (North) Forest Reserve at Bukit Cheeding (Figure 6.2) ever the choice of the Village Action Committee. The site near Perang Besar proposed in the EIA was rejected apparently because it was already Orang Asli Reserve land, but belonging to a different community.[54] Batin Senin Anak Awi, the headman of Kampung Busut Baru ('New' Kampung Busut) reported that the community did not even see the Bukit Cheeding site before resettlement took place.[55] His assessment of the new site contrasted sharply with K. L. International Airport Berhad's official video documentary which considered that 'the exchange is more than fair for people who once lived only on the forest' (K. L. International Airport Bhd, 1994a). The land around the village was cleared when they arrived with the promise that newly-planted oil palm would be mature 'when the first plane takes off'.[56] After four years – and with less than a year until the opening of the airport – planting had still not begun (see Figure 6.4). According to the Batin, this was because much of the land in the area was swampy and, therefore, not sufficiently firm for oil palm.[57] In Kampung Busut Lama, Batin Senin had worked some 20 acres (around 8 hectares) of oil palm and rubber which gave him a monthly income of around RM 2,000. Many villagers, he said, had lost income because of the unsuitability of the peat at Bukit Cheeding for agriculture, although some had been able to find work at the nearby Boh tea estate and in local 'Chinese' factories. Resettlement was thus more a process of proletarianisation than one through which a new generation would be able (as the official video put it) 'to enjoy the brighter future promised by Malaysia's Vision 2020' (K. L. International Airport Bhd, 1994a) – and less still a process sensitive to the environmental specificity of Temuan cultural practices.[58]

To the extent that there was authoritative recognition of the 'culture' of the Temuan – or, rather, of the 'Orang Asli' – at KLIA, this only served to

Figure 6.4 Awaiting the take off of Kampung Busut Baru.

Source: Author's photograph, 1997.

legitimise their socio-spatial marginalisation. On the one hand, as we noted in the previous section, the official aim of the state – and the JOA in particular – towards Orang Asli has been one of infrastructural modernisation. This was clearly the case at the Bukit Cheeding resettlement site for Kampung Busut, even if there were chronic problems of implementation. Official figures suggested that RM 4.6 million in compensation was paid to the 83 Orang Asli families at KLIA and a further RM 8.8 million spent on the construction of houses, a school, a *surau*, road and other infrastructure,[59] although the new site was not supplied with electricity until 1995 and did not receive piped water until 1997.[60] In addition, according to the Batin at Kampung Busut Baru, the poorly designed and constructed houses made them noisy during the rain and constantly hot. While each house was said to have cost RM 25,000, Batin Senin calculated their value at less than half that amount. On the other hand, a tendency to 'traditionalise' Orang Asli 'explained' and justified such developmental failure. As a family is shown leaving Kampung Busut Lama in the official video, the narrator comments that, 'their houses seem already to have returned to the forest'. Later, the resettlement site houses are said to be 'in the traditional style which they prefer' (K.L. International Airport Bhd, 1994a).[61] Orang Asli cultural preferences here, then, are portrayed as little more than a 'traditional' landscape past left behind by more 'developed' others on the path to modernity.[62] Such traditionalisation of Orang Asli has not only justified their displacement to make way for infrastructural modernisation, but also served to consign them to peripheral/'traditional' spaces of (post-)colonial underdevelopment (see also Nicholas, 1989).

Failure to acknowledge the potential of indigenous knowledges as resources for development was evident from plans to make land adjacent to the resettlement site into a wetland sanctuary. As construction work on KLIA proceeded in 1996 – and as Kampung Busut Baru awaited the installation of piped water – a 2,000 hectare site was made available by the Selangor state government to 'showcase environmentally sound development methods' and provide 'a small working example of sustainable development' (Tan, 1996: 7).[63] No mention of the presence of Orang Asli was made in the newspaper coverage despite the fact that an accompanying schematic map located the sanctuary on the parcel of land directly south of Bukit Cheeding Tea Estate occupied by the Temuan from KLIA. At one level, the site's 'global' wetland potential confirmed what residents of Kampung Busut Baru already knew: this land could not sustain oil palm cultivation. In addition, however, it was sadly ironic that the now cartographically invisible Temuan had been moved to the Bukit Cheeding area from their ancestral territories during the UN's International Year for the World's Indigenous Peoples (1993) which recognised their 'special relationship' with the land (Nicholas, 1993: 1). While Kampung Busut Baru was (not) seen as part of a 'global focal point for sustainable development' in line with Malaysia's obligations under Agenda 21 of the Rio Summit and the Ramsar Convention (Tan, 1996: 1), Kampung Busut (Lama) had become a runway (Runway 1) at KLIA. And while the UN recognised indigenous peoples as 'models of economic sustainability' (Razha, 1995: 1), the resettlement of another Temuan community at Bukit Lanjan, west of Kuala Lumpur, led one Malaysian reporter to question whether this would be the last generation to 'lead a life that is still in tune with nature' (Yusof, 1997: 4).

Nonetheless, the proposed wetlands project diagnosed a broader valorisation of 'nature' which was also evident in the design of KLIA: 'Airport in the Forest; Forest in the Airport' (K.L. International Airport Bhd, 1994b). This promotional descriptor referred to the fact that the main terminal building was to be 'surrounded by landscaping zones through which various departure and arrival routes will pass', while the satellite building was designed to contain a tropical rainforest (K. L. International Airport Bhd, 1996: np). The valuing of nature of which such designs are symptomatic may be understood in two ways. The first is simply economic or commercial value. Writing in the epilogue to an edited collection on 'The Indigenous Minorities of Peninsula Malaysia', Hood Salleh described how the aestheticisation and commodification of the environment had created only 'little pockets of greenery, spots of beauty and tranquillity, little niches here and there, small gardens for escape away from the great urban garbage' (Hood, 1995: 131).[64] A second and related valorisation concerns conceptions of proximity to nature as a vital component of healthy, modern living. We have seen in Chapter 5 how supposedly harmonious relations with nature were central to MSC governmental ideals. The design of KLIA was explained by its Japanese architect in terms of a 'theory of symbiosis' which included conceptions of a mutually beneficial coexistence of people and nature (Kisho Kurokawa, cited in Peter, 1997: 2). Kurokawa's project thus elaborated the value of nature to intelligent futures: 'as much as possible of the natural environment is to be preserved in a fusion of technology and nature' (Kuala Lumpur International Airport, 1994a).

Kurokawa's ideas also gave architectural and landscape form to discourses of 'Asia' and the need for an 'Asian way' to modernity. These included, but extended beyond, proximity to nature as 'an integral part of Asian identity' (Alina, 1997: 11): 'What I am proposing in the region is a new intrinsically Asian theory of urban planning rather than merely importing ideas from the West', Kurokawa said in an interview in Malaysia (cited in Peter, 1997: 2). At the *South-South Mayors Conference*, where we began this chapter, then Deputy Prime Minister, Anwar Ibrahim, highlighted the urgency of finding such an Asian urbanising route, in opposition to cities of the West (or, rather, 'the North'):

> In some cities of the North, the dichotomy between rich and poor have (*sic*) solidified into a permanent divide, creating a new apartheid. The rich cluster themselves in Suburbia where the marketplace supplies the best amenities: schools, healthcare and recreation while the have-nots gravitate into the decaying inner cities, trapped in a perpetual cycle of poverty, illegitimacy, single parenthood, delinquency, educational under-performance, drugs and crime.
>
> (Anwar, 1997: np)

These were certainly social ills and spatial divisions that no Malaysian would wish to import. Kurokawa's Asian vision thus appeared to be a fitting rhetorical end to Route 2020. Yet, more critically, the implications of KLIA for the Temuan from Kampung Busut Lama diagnosed socio-spatial dividing practices associated with the discourses and landscapes of 'Asian' modernity.

In this chapter, I have considered socio-spatial divisions associated with the development of Kuala Lumpur Metropolitan Area in the 1990s. These divisions were partly an extension of existing processes of economic transformation. A legacy of colonial land laws placed communities and classes in differentiated ways in relation to spatial processes of development. In addition, increasingly private-sector-led development, in Malaysia as elsewhere, has done little to ameliorate the lives of the most socially marginal groups. However, I have also suggested that it is possible to identify new socio-spatial dividing practices in 1990s Malaysia associated with rationalities of 'intelligent' development. The apparent urgency to produce appropriate sites and transport connections for intelligent futures had spatial consequences deserving of critical geographical scrutiny. I have shown how intelligent discourses were actively bound up in material practices of social and spatial marginalisation.

For all the attempts to differentiate landscapes of 'Asian' modernity, Route 2020 led to some uncannily familiar outcomes. The urban planning and design of Kisho Kurokawa's KLIA at the southern end of the MSC chimed in harmoniously with the development discourses of city authorities and federal political elites. Deputy Prime Minister, Anwar Ibrahim, warned against the social and spatial problems of urbanisation in the North. Yet Malaysia's own intelligent urban utopias suggested similar dividing practices: not only a planned suburban shift of middle classes long evident in North America and elsewhere; but also more recent spatial splintering associated with informational forms of society and economy. MSC development centred upon interconnected intelligent cities and citizens while (further) peripheralising plantation workers and indigenous groups. Yet the variegated zones of government sketched by Aihwa Ong imply more than just where different groups ended up (Ong, 1999). Rather, space was actively bound up with the increasingly differentiated government of the population. While 'world class' knowledge workers and suitably 'learning' Malaysians were free to realise themselves in intelligent ways, indigenous citizens were consigned to a traditionalised 'aboriginal periphery' under the pastoral guidance of JHEOA. Ong's use of 'aboriginal' here is particularly apt suggesting, as it does, the way in which traces of colonial demarcations of society and space inform unfolding geographies of inequality. Shifting constellations of uneven geographical interconnection yielded newly-differentiated experiences of modernity in the Malaysian present.

7 Conclusion

To write, as I have done in this book, about the modernity of a particular geo-historical location is not to produce a human geography without wider scope or application. In the first section of this concluding chapter, I seek to highlight the geographical contribution of this analysis of 'intelligent landscapes' in late 1990s Malaysia. In the first place – and as has been outlined in Chapter 2 – the very conceptualisation of modernity in this research is an intervention into a contested intellectual terrain. The study of the modernity of particular places clearly heeds renewed calls for geographical specificity in social science research (McGee, 2002). Yet I have also sought to avoid conceptions of bounded ('Malaysian' or 'Southeast Asian') 'alternative' modernities. Rather, I have reconceptualised difference in terms of a networked spatiality of asymmetrical interconnection. On the one hand, therefore, transformation detailed in the preceding chapters has not been reducible to trends diffusing from supposedly more 'advanced' people and places (typically, the so-called 'West')[1] and, in turn, findings here can not in any simple sense be used to 'explain' or to account for socio-spatial change elsewhere. On the other hand, the landscapes analysed 'On Route 2020' – the sites and the way of seeing that they produce and which produce them – connect up to all kinds of other(s') places. The critical geographies of this book have relational resonances in diverse geo-historical locations.

Geographical connections

There are three broad geographical dimensions of this research which speak to transformations in other(s') places. The first concerns *the role of urban space in positioning nation-states in 'global' cultural-economic networks.* I have shown how MSC was conceived as a 'wired' and interconnected urban zone conducive for high-tech investment, especially at Cyberjaya (Chapter 5). The siting of the new corridor as a southward extension of Malaysia's existing main metropolitan centre was not accidental. Even prior to the launch of MSC, the federal government had assumed progressively greater control over – and had increasingly focused infrastructural investment in and around – the national capital, Kuala Lumpur (Chapter 4). In addition, work was already under way on a new international airport at Sepang – making known a new 'southern corridor' of greater Kuala Lumpur – which was intended to propel the urban region to the status of transport 'hub' for Southeast Asia. New transport links, including the appropriately nick-named 'Route 2020', were planned to integrate the southern corridor as part of an extended Kuala Lumpur Metropolitan Area (KLMA) (Chapter 6). The decision to locate the new federal government administrative centre, Putrajaya, within this zone has a significance beyond its role in giving coherence to the new corridor. The Selangor state's hand-over of control of the 4,581 hectare territory to the federal government on 1 February 2001 was further evidence of a 'rescaling' of central government power and control to the nation-state's would-be global 'node' (see also Bunnell, 2002b).

Much of the work of (re)positioning was symbolic or imaginative. On their road show of North America and Europe in 1997, the Malaysian Prime Minister and the Multimedia Development Corporation elaborated MSC as a 'greenfield' urban 'test-bed' free from constraints or vested interests and thus as a suitable site for the regional operations of 'world class' IT companies (Chapter 3). Projection of the existing city through the rapidly-rising Kuala Lumpur City Centre (KLCC) had already begun to revise upwards international imaginings of the city and nation. The Petronas Twin Towers' record-breaking height conferred 'KL' with unprecedented global visibility while the media 'travels' of the building reimagined the Orientalist urban landscape as an 'investible', modern metropolis. The 'national' prioritisation of this post-colonial revisioning in Malaysia was such that, in 1999, Mahathir himself rebuked the Hollywood film-makers of *Entrapment* for splicing the Petronas Towers alongside riverside slums shot in Malacca (see Bunnell, 2004). As for cities elsewhere in the 'dynamic' Asia-Pacific of the 1990s, the changing urban landscape was the most visible sign of city authorities' attempts at global cultural as well as economic repositioning.

Rather than merely considering urban landscapes as representing processes of national repositioning, however, I have detailed how MSC space performed an active role in geo-historical change. Thus – and this is the second broad geographical dimension of the research which has resonances beyond MSC-Malaysia – I have exemplified *the 'work' of landscape in (re)forming ways of seeing and being*. This was most clearly evident in Chapter 4 where the sheer size of a single landscape artefact, the Petronas Towers, ensured its centrality to everyday views on/of Kuala Lumpur. The building's mediated visibility meant that it was also enrolled into geographical subjectivities beyond the physical limits of the city. KLCC's proponents framed the Petronas Towers' record-breaking height in terms of national 'progress' and 'development'. Yet as everyone came to have a view on the building, I have also shown how it became a symbolic focus for critical perspectives on urban and national development. Nonetheless, the normative landscaping of which KLCC formed part shaped not only how citizens saw the(ir) city, but also how they saw themselves and their own conduct in relation to increasingly 'world class' standards.

In a rather different way, I have 'read' plans and political projections for new city spaces in terms of their intended citizen-subject outcomes. While intended to fulfil rather different functions, the two new MSC cities, Cyberjaya and Putrajaya, were founded on a modernist governmental faith in the possibility of creating a new generation of 'intelligent citizens'. 'Intelligent' here extended beyond the infrastructural wiring of urban space, referring more broadly to ways of living and working which were deemed appropriate for national 'development' goals. In Cyberjaya, this meant the provision of spaces of physical and electronic interconnection which were 'conducive for creativity', but which also averted the apparent 'dark side' of leading-edge technopoles elsewhere (in 'the West'). In Putrajaya, emphasis was on 'indigenous' norms and forms as cultural resources for a specifically 'Malaysian' urbanity. Particularly in Putrajaya, 'Western' contributions to intelligent environments were downplayed, but both MSC cities have geographically-extensive geneaologies (Chapter 5). I have also shown how the landscape work of Cyberjaya and Putrajaya – their governmental role in fostering new forms of conduct and ways of seeing – preceded their material completion or habitation: intelligent city sights were enfolded into citizens' systems of evaluation through technologies of visualisation and electronic interaction at public displays and exhibitions.

While exhibitions and official descriptions framed MSC in terms of a national route to development into which all Malaysia(ns) could be incorporated, I have highlighted new geographies of exclusion. Put another way, the third geographical dimension of this research has been to reveal *socio-spatial dividing practices associated with ostensibly 'alternative' (non-Western) routes to 'development'*. If KLCC projected Malaysia closer to its official goal of 'fully developed' status by the year 2020, it was precisely by setting new 'global' standards for the city – standards and systems of evaluation which marginalised certain places and people in/from globalising KL (Chapter 4). The latest *Draft Structure Plan* for 'Kuala Lumpur 2020' describes the presence of squatters in the city's 'world class' landscape as simply 'unacceptable' (City Hall Kuala Lumpur, 2003: 12–13). Similarly, would-be 'intelligent' MSC city futures were founded on the displacement and exclusion of already socio-spatially marginal groups such as plantation workers (Chapter 6). The colonial resonances of moral geographies of 'intelligence' here are highly significant (and demand further research). Malaysia's putatively non-Western route to development mapped people and places along a teleological route to a modern 2020 led by supposedly intelligent cities and citizens. This singular modernising trajectory (dis)placed Orang Asli as 'traditional' laggards rather than allowing for the possibility of culturally-specific alternative modernities (see also Benjamin and Chou, 2002).

Critical geographies of MSC, then, extend beyond a simple dichotomy of inclusion in versus exclusion from intelligent spaces. Plantation workers displaced from the sites of the MSC cities *have* been incorporated into landscapes of Malaysia's information economy – as security guards, gardeners and cleaners (see also Bunnell, 2002a). As I have shown (in Chapter 2), Aihwa Ong's concept of 'graduated citizenship' is a useful way of thinking about how populations are increasingly bio-politically segmented in the way in which they relate to national (and transnational) projects. I have shown how MSC may be understood as a distinct space of government. Foreign knowledge workers and would-be intelligent citizens were exempted from many of the political and socio-cultural – as well as economic – regulations operating in other parts of the national territory and so were 'free' to realise themselves in creative, innovative ways (Chapter 5). This, as I have shown (in Chapter 6), contrasted sharply with displaced Orang Asli groups from the KLIA site: resettled to an 'aboriginal periphery' where their 'development' is monitored and regulated by a (post-)colonial administrative department. More stark still, have been modes of power associated with the regulation of foreign (non-knowledge) workers. The 'unskilled' bodies which carried out much of the physical labour of constructing 'world class' Malaysian national landscapes during the rapid growth period of the 1990s were increasingly subjected to detention and expulsion following the onset of the Asian financial crisis in July 1997 (Rajaram and Grundy-Warr, 2003). 'Illegal' and/or 'foreign' workers – the frequent discursive elision of the two is significant – were rendered morally as well as economically out of place. The detention camp became perhaps the bio-political 'other' space to the MSC liberal zone of government.

Economic and political events which began in 1997, of course, had much wider implications for the positioning of MSC and Malaysia. The introduction of capital controls in September ran against the neo-liberal orthodoxy of supranational financial institutions such as the International Monetary Fund. However, it was perhaps Mahathir's discursive responses as much as policy practices that tarnished the image of 'liberal' investibility established in and through MSC. If notions of MSC as a liberal 'test-bed' had allowed mediatory management of 'global' and national political economic prerogatives (Chapter 3), the financial downturn from mid-1997 saw Mahathir seeking to retain domestic political legitimacy by blaming proponents and processes of neo-liberal globalisation. In speeches on the causes of the crisis, for example, Mahathir (1998b) railed against the 'religion' of the free market for punishing those who do not 'believe' or 'practice' (see also Kelly, 2001); he was demonised in the eyes of the international financial community as the regional 'bad boy' (Jomo, 2001: 14).[2] It should also be noted that this was an imagined 'region' that had lost its 'miraculous' allure (Fukuyama, 1998). History and geography, so it seemed, were no longer on Asia's side: as Malaysia recovered economically, association with a 'sick' region became an imaginative cartographic burden. The 'contagion' of economic developments beginning with the devaluation of the Thai baht on 2 July 1997 was clearly more than just a temporary dent in the self-confidence that had emboldened modernising self-theorisation and projected a seemingly clear path to 2020 (see Hilley, 2001). Yet while such change was cast both domestically and internationally in terms of 'problems' under the generalised label of 'crisis', it also presented (and continues to present) new possibilities. It is with some reflection on alternative ways forward which I end this book.

New ways forward?

It was escalation 'from economic to political crisis' (Gomez and Jomo, 1999: 185) that brought to prominence new possibilities for alternative conceptions of and routes to development. Differences of opinion between Mahathir and his deputy as to how the financial situation should be managed ultimately led to Anwar Ibrahim's ousting from government. His dismissal, arrest and prosecution for corruption and sexual misconduct fomented a new alignment of social and political opposition to *Barisan Nasional* (*BN*) and Mahathir in particular (see Sabri, 2000). Events which began as '*Reformasi*' in protest at the treatment of Anwar Ibrahim grew into a new electoral opposition coalition, the *Barisan Alternatif* (*BA* or 'Alternative Front') in 1999.[3] At least three aspects of this transformation suggested a 'new politics in Malaysia' (see Loh and Saravanamuttu, 2003). First, the 'Anwar Affair', unlike the political 'crisis' of the 1980s *within* UMNO, divided the Malay community in such a way as to hint at an 'unravelling' of Malay political unity (Maznah, 2001). Second, Malay middle classes who had previously supported *BN* were involved for the first time in large numbers in political opposition and demands for democratic reform (Saravanamuttu, 2003). Third, the diversity of political parties and social voices stirred into this ferment of 'opposition' suggested electoral contestation beyond the established 'post-colonial' politics of ethnicity. While the *BA* was unable to prevent the government from securing a two-thirds majority at the election of November 1999, *BN's* win was secured by more than merely playing its ethnic cards right. Apart from the first-past-the-post electoral system which gave *BN* more than three-quarters of the parliamentary seats despite winning only 56.5 per cent of the vote, the ruling coalition was also assisted by an economic upturn during 1999. Economic recovery translated into electoral success in a 'political culture' of 'developmentalism' which associates *BN* with growth, consumer affluence and the 'political stability' upon which such a route to modernity was supposedly founded (Loh, 2003: 261).

The MSC not only became a series of high-profile sites/sights in electoral contest, but was also bound up with the opening of new domains for social and political contestation. On the one hand, while Anwar Ibrahim, as Finance Minister, shelved or postponed a number of megaprojects – including transport infrastructure for MSC such as 'Route 2020' – the post-Anwar National Economic Recovery Plan (NERP) stressed that MSC was 'the next engine of growth and is fully backed by the Government' (Multimedia Development Corporation, 1999). Mahathir's reluctance to cut back large-scale infrastructure development projects, including in the MSC, was cast in international and domestic opposition media as imprudent attachment to 'pet' schemes (Sardar, 1998). Putrajaya, in particular, became the metasymbol of a state capitalism synonymous with self-legitimation, lavish monumentality and lack of transparency (Maznah, 2000). Petronas, the national oil company of twin towers fame (Chapter 4), was involved in controversial acquisitions[4] and continued to bankroll the federal government administrative centre[5] with its 'palatial' Prime Minister's residence: all further fodder for Mahathir's multiplying detractors. On the other hand, the state-led high-tech transformation of which MSC formed part had also spurned new electronic spaces for the dissemination of 'critical' perspectives. Even prior to the crisis, websites were proliferating as important sources of alternative news and views in the context of a state-owned and regulated media. Expansion was boosted by both the news- and gossip-hunger generated by the controversial Anwar Affair and subsequent electoral vying in 1999. In the lead up to the November election, moments of political openness in the mainstream press (see Chapter 6) were greatly diminished (Mustafa, 2003). It was in no small measure MSC's investment-friendly commitment to not censoring the Internet that provided an environment conducive for intelligent new spaces of contest to dominant conceptions of 'development' (including, but certainly not limited to, critique of the MSC itself).[6]

There has been increased awareness and even official acknowledgement of the limitations of the MSC. While official support for the project never waned, the cities at its 'core', particularly Cyberjaya, suffered inevitable delays to the development of supporting infrastructure after 1997. One can only imagine that the marketing team behind 'Cyberia' must have been blissfully ignorant of potential for linguistic conflation with Soviet-style banishment to hostile environments given the barren landscapes surrounding a nascent 'dot-com property in Cyberjaya' in 1999.[7] In these conditions, it was no surprise that existing companies and suitably skilled workers were reportedly reluctant to move to MSC (Mellor, 2001). Yet MSC also appeared to suffer from a diminished investibility at the nation-state scale: although MSC status companies were exempted from 'national' capital controls, there was a marked reduction in investment in the corridor (Indergaard, 2003). According to perhaps the most high-profile critique of MSC – in *Business Week* magazine in March 1999 – this predicament was a result of 'Mahathir's high-tech folly': 'Mahathir's powerful vision helped make the MSC a reality. Now his actions threaten the future of his dream' (Einhorn and Prasso,

1999: 22). While in the context of an imminent election, flat denial of such appraisal was inevitable, more recently, Mahathir has acknowledged MSC's failure to live up to expectations (Indergaard, 2003). A broader dot-com 'bust' has not only corrected inflated expectations of high-tech economic returns, but also meant that 'failure' could be attributed to 'external' conditions. While the technological utopianism of the mid-1990s has certainly faded, however, some now consider that MSC is actually faring well when compared to other digital districts such as New York's 'Silicon Alley' (see Indergaard, 2003).[8] By the infrastructural and economic criteria of its proponents, MSC is perhaps a qualified success.

Yet, as I have sought to show throughout this book, it is important to problematise high-tech 'success' in terms of broader social and spatial implications. While supposedly 'critical' journalists and political opponents have found abundant ammunition with which to attack MSC in its own terms – highlighting discrepancies between boosterist 'planned' and actual completion schedules, investment targets or innovation achievements – what are the broader presumptions and effects of globally-oriented 'intelligent' progress? More specifically, to return to a theme running through this research, to what extent can an apparent shift away from ethnic-centred national development be considered a progressive way forward? On the one hand, efforts to 'woo' knowledge workers to MSC and Malaysia more broadly have been extended. In his 2001 Budget Speech, Mahathir declared that:

> To ensure success from the new economy, we need a pool of the best talents from at home and abroad. Efforts need to be undertaken to hire the best brains regardless of race and nationality, from Bangalore to California. This is a step towards creating a world-class workforce.
>
> (Ministry of Finance Malaysia, 2000)

It is 'Asian achievers' who have been singled out to enjoy the privileges of 'right to abode' or even citizenship as 'world class foreigners' (Ministry of Finance Malaysia, 2002: 48). To the extent that a large proportion of the beneficiaries of such labour liberalisation are 'non-Malays' – the *K-economy Masterplan* specifically identifies 'East Asians' and 'South Asians' – so-called 'knowledge economy' measures might be welcomed as part of a move to less ethnically-asymmetrical means to 'national' economic development. Yet, on the other hand, I would caution against such an over-generalised rosy reading. While the latest ('post-NEP') phase of Malaysia's political economy indeed appears to run against any narrow defence of 'Malay interests', the beneficiaries of actual and proposed labour and immigration enticements in Malaysia, as elsewhere, are a highly-skilled professional elite (Cheng and Yang, 1998). What are the systems of evaluation governing differential bio-political incorporation in – and even exclusion from – a 'global' information economy in Malaysia? While 'world class' professionals are wooed to

Malaysia and have the right to fair treatment in their new place of residence 'regardless of race and nationality', for workers of other 'classes', harassment and displacement are increasingly normalised, irrespective of their (post-)colonial ethnic/racial ascription (Bunnell, 2003). And, as we saw in Chapter 6, it has not only been 'foreign workers' who have borne the costs of 'global' development along Route 2020.

Finally, however, recognition of increasingly transnational interconnection and contestation is perhaps one way towards imagining 'new politics'. In Chapter 6 we noted that while Temuan land at Bukit Tampoi was appropriated for the transport globalisation of KLMA, Orang Asli have also been enrolled in transnational networks of 'indigenous peoples'. 'Global' experiences and interconnections were mobilised in making the case for the Temuan at Bukit Tampoi.[9] In however small a way, the ruling of April 2002 in favour of Sagong Tasi and the other plaintiffs (Sagong bin Tasi & Ors v Kerajaan Negeri Selangor, 2002) showed how 'global forces' – conventionally imagined to be aligned with mobile capital(ists) and powerful institutions – can effect progressive shifts in the placing of socio-spatially marginal groups. This book as a whole is founded on a recognition of the difference that connections make – as distinct from bounded topologies of 'otherness'. Bringing geo-historical specificity into 'modernity' not only attends to spatialised difference, but also allows critical interrogation of presumptions and constructions of difference in particular times and places. Farish A. Noor's brilliant writing – originally disseminated through the very channels opened by Mahathir's push for high-tech, global connectivity[10] – has shown how 'elements of the past that have been relegated to the margins or footnotes of political history' are resources for scripting alternative social and political paths (Farish, 2002: 2). Might 'the Other Malaysia', in a modest way, be supplemented by critical *geographies* which unmake taken-for-granted 'intelligent' landscapes – material spaces and the ways of seeing which form, and are (re)formed by, them? Thinking about *where* the present came from unsettles the essentialist ground of elite 'Asian' or 'Malaysian' modernities. This is not just a matter of opposing Mahathir's or BN's grand vision, but about problematising the intertwined routes and roots (see Clifford, 1997) of a geo-historically specific political rationality.

Notes

1 Introduction

1 The IAP first met at Stanford University on 16 January 1997. According to the official MSC website, 'It was the first time that such a distinguished group of international business leaders had assembled in one room to provide counsel to a national government on creating the ideal technology environment'. What is more, 'Like the MSC, the IAP meeting was truly a world-first' (Multimedia Development Corporation, 2003). More detail on the International Advisory Panel is also provided in Chapter 5.
2 During the eight years from 1988, the Malaysian economy grew by an average of 8.9 per cent per annum. The *Sixth Malaysia Plan* period, 1991–5, was described as 'a momentous period of rapid progress', with an average annual increase of 8.7 per cent (Malaysia, 1996).

2 Modernity, space and the government of landscape

1 On the global scale power of development discourse see Escobar (1995).
2 Or, as work in anthropology has suggested, there is a need to 'rethink difference through connection' (Gupta and Ferguson, 1992: 8).
3 Similarly, Derek Gregory's critique of Marshall Berman is concerned precisely with a perception that the latter's understanding of capitalist processes yields 'an undifferentiated, almost timeless representation of modernity' (Gregory, 1994: 293).
4 This is what Appadurai has aptly referred to as the 'spatial incarceration of the native' (Appadurai, 1988).
5 For Moore, this would necessitate a dialectical conception of the global and the local. This involves seeing the global not as a monolithic whole, but as 'a set of situated and interrelated knowledges and practices, all of which are simultaneously global and local' (Moore, 1996: 9).
6 Though 'the region' itself, of course, is an 'imagined' territory (see Sidaway, 2002). The imaginative work of region building is particularly apparent in 'Southeast Asia' which, until the 1960s at least, frequently defied cartographic legibility, 'often found straddling maps on different pages' (Cairns, 2002: 110). On the intertwined intellectual and geopolitical construction of Southeast Asia, see Anderson (1998).
7 For a consideration of other typologies of the state in Southeast Asia, see Malhotra (2002).
8 See Dean (1999) and Rose (1999) for excellent overviews.
9 More specifically, the recent literature on governmentality has focused on the English-speaking world (see Rose, 1999; though see also Sigley, 1996 for work on East Asia).

10 See, for example, Abidin Kusno's work on architecture and political culture in Indonesia. As Kusno shows, even in the colonial and New Order regimes, political power included 'ethical' strategies to foster the self-regulating capacities of the population (Kusno, 2000).

11 For critical reflection on the limitations of Anglo-American urban theory for understanding urbanisation and globalisation in Southeast Asia, see McGee (2002).

12 This is to be differentiated from a wider literature on 'post-development' (see Watts, 2003).

13 Ong identifies at least six zones of graduated sovereignty for the population as a whole (Ong, 1999: 218).

14 It is important to distinguish this sub-disciplinary use of the concept from the way in which 'landscape' has been deployed in other geographical work. See, for example, Philip F. Kelly's 'Landscapes of Globalization' on the political economy of late twentieth century Philippines (Kelly, 2000).

15 See Thrift (1996b) for the most comprehensive treatment of these developments in human geography.

16 On the material as well as symbolic role of urban space in fostering new social and political subjectivities, see Kusno (2000).

17 On the way in which authoritative aims and objectives which seek to shape and normalise subjects through 'action at a distance' see Rose and Miller (1992).

3 Positioning Malaysia

1 'Race', rather than ethnicity, captures the biological determinism of British imperial categorisation.

2 I use '(post-)colonial' here to problematise conceptions of a specific historical moment – typically marked by a change in political status – as delimiting the 'end' of the 'colonial'. For a recent critical review of postcolonialism in geography see Yeoh, B. (2001).

3 Brunei declined principally because the Sultan was not satisfied with the financial arrangements for his oil-rich state.

4 With the important exceptions of the Malay aristocracy and some very wealthy non-European individuals.

5 Each of the nine Malay states was ruled by a sultanate. By the end of the nineteenth century, Perak, Selangor, Negri Sembilan and Pahang had accepted British Residents and were collectively termed the Federated Malay States (FMS). The remaining states of present-day Peninsula Malaysia – Terengganu, Perlis, Kedah, Kelantan and Johor – were later induced to accept British 'Advisers' (as opposed to Residents) and these became known as the Unfederated Malay States (UMS). Together with the Straits Settlements, the FMS and UMS formed British Malaya (Comber, 1983). Muhammad Ikmal Said (1996) shows how the British increased the prestige of the Malay rulers through the 'restoration' or 'invention' of 'traditional' protocol for their installation and in other state ceremonies.

6 A number of plantation companies are involved in MSC, not only as owners of much of the land on which it is being built, but also in various aspects of construction and development. These non-agricultural roles are evidence of their diversification. See Chapter 6.

7 The special position referred to privileges in the public service, Malay land reservations, scholarships and educational grants and quotas for licences and permits (Ong, 1990).

8 Muhammad Ikmal shows how, rather than 'unmasking the "fiction" of British protection, UMNO, and Malay opinion generally, adopted that "serviceable" but fictionalised past as a real one' (Muhammad Ikmal, 1996: 51).

9 'The struggle was called an "Emergency", rather than a war, because an officially designated war would have sent the insurance premiums for Malayan rubber and tin soaring' (Ascherson, 1998: 3).

10 Despite winning twenty three fewer seats than in the previous election of 1964, the Alliance still retained enough seats to form a majority in parliament. Nonetheless, gains made by the predominantly-Chinese Democratic Action Party (DAP, which opposed what it saw as racial hegemony and supported the earlier PAP 'Malaysian Malaysia' concept), Gerakan (a non-communal party) and the People's Progressive Party (PPP, which championed Chinese rights in Perak), the three of which had entered into an electoral pact, meant the prospect of a strong Chinese-based opposition in parliament for the first time (see Goh, 1971).

11 The remainder of this section draws upon Comber (1983).

12 *Rukun* means 'like the ideal relationship of friendship' or 'without quarrel'; *negara* means 'nation' (Comber, 1983).

13 As noted earlier, *bumiputera* is the constitutional term given to Malays and other 'indigenous' groups particularly in East Malaysia.

14 In addition to expanded educational efforts which led to the formation of a Malay middle class, redistribution was to be achieved by the creation of public or state-owned enterprises (SOEs). Gomez and Jomo (1997) divide these into three categories: departmental enterprises largely concerned with public services; government-owned private or public listed companies, which include *Bumiputera* trust agencies (such as *Perbadanan Nasional Bhd.* ('National Corporation', Pernas) and *Permodalan Nasional Bhd.* ('National Equity Corporation', PNB) for the accumulation of wealth on behalf of the whole community; and statutory bodies established by law at federal and state levels. The last of these includes the state oil company, *Petroliam Nasional Bhd.* (Petronas) and it is significant that the export of petroleum from Malaysia from the mid-1970s enabled increased public spending without dramatic increases in foreign borrowing.

15 Continued economic growth was based on export-oriented industrialisation (Gomez and Jomo, 1997). Thus, at the same as Malaysia moved towards greater state regulation of many aspects of the economy, a Free-Trade Zone Act (in 1971) and an Investment Incentives Act (in 1972) provided for a system of tax exemptions and a strong legislative framework to attract internationally footloose industries. The development of free-trade zones in Penang has been the subject of geographical research (see Eyre and Dwyer, 1996).

16 In addition, a number of researchers have noted the gradual integration of Chinese and *bumiputera* capital (e.g. Searle, 1999). Integration, particularly with politically-connected Malays, forms one half of an often-cited two-pronged strategy of Chinese business elites – the other being the forging of overseas Chinese alliances and networks (Heng, 1992).

17 A term used by Lee Kuan Yew, leader of the People's Action Party (PAP) and subsequently Prime Minister of Singapore, to describe those politicians and intellectuals who vehemently opposed Chinese economic domination and/or political encroachment on the 'special position' of the Malays (Khoo, 1995).

18 The colonial influence upon Mahathirist thinking is discussed below.

19 Mahathir's own father, Mohamad Iskander was the son of an Indian Muslim and a Malay woman. It is this ancestry which is often used to 'explain' Mahathir's aggressive 'un-Malay' political style (Morais, 1982)

20 Medical metaphors are often used in relation to Mahathir's ideas. An article in *Time*, for example, described him as a politician who 'wields his notebook as on on-going prescription pad for Malaysia's ills' (Spaeth, 1996: 4). It would seem that Mahathir himself considers his medical training in relation to the practice

of his politics: 'If the ailments of a society or nation are attended in the same way as the illness of a patient, some good results must follow. The essential thing is to develop good diagnostic skills' (cited in Adshead, 1989: 53).

21 Khoo Boo Tiek (1995) describes Mahathir's Social Darwinism as embodying a paradox between Mahathir's explanations for Malay debilitation, on the one hand, and protectionism as a means to overcome these on the other. David Livingstone, however, has shown how Lamarckism could be 'mobilised to justify the politics of interventionism' (Livingstone, 1992: 189).

22 Hamzah in fact noted an increase in the Malay population of Kuala Lumpur before NEP. This was in part diagnostic of the very urban dominance that he described.

23 Though this was clearly also sustained through connections with events elsewhere in the 'Muslim World', particularly following the Arab-Israel War of 1973 (Hussin, 1993: 13).

24 The original Malay version, '*Menghadapi Cabaran*' was published by Pustaka Antara in 1976.

25 Mahathir sought to augment his own and UMNO's Islamic credentials through moves such as the establishment of an International Islamic University and an Islamic Bank and accelerating the government's mosque-building programme (Camroux, 1994).

26 It is perhaps for this reason that, from 1983, Mahathir attempted to curb the hereditary powers of the Sultans. As Rehman Rashid (1993: 191) has put it: 'The sultanates, by their very existence as an institution, restrained the upliftment of the Malays. They bid them genuflect; they kept them humble. (I imagine this must have made Mahathir's blood boil.)'

27 Khoo Boo Tiek notes a transformation in Mahathir's nationalism between *The Malay Dilemma* and *The Challenge*. In remarkable contrast with the former, the word 'Chinese' is not even mentioned in *The Challenge* until five pages from the end of the book. Mahathir's Malay nationalism was thus refocused 'away from the divided state of Malaysia to the divided states of the world' (Khoo, 1995: 42).

28 This is often associated with the Institute of Strategic and International Studies (ISIS) founded in 1983. ISIS Director, Noordin Sopiee is thought to have played a leading role in the proposals for an East Asian Economic Grouping (later 'Caucus') and some consider his influence to extend beyond foreign policy.

29 A broad 'third way' was defined between the command model, on the one side, and pure market capitalism on the other. Malaysia Incorporated and the Heavy Industries campaigns – based respectively on Japanese and South Korean precedents – promoted a strong state role in the organisation of big business which was conventionally regarded as a key criterion of an Asian model (see, for example, Wade, 1990). In Malaysia and elsewhere, this has been referred to as 'state capitalism'.

30 According to Khoo Boo Tiek, Mahathir's father was from that 'breed' of colonial school teachers in Malaya who enforced not only English authority and discipline, but also worthy values, personal habits and moral responsibility to a new generation of aspiring professionals and administrators. There is no reason to doubt that both aspects extended beyond the classroom and into the family home. Paternalistic lessons of hard work and moral restraint were meant to last a lifetime; they inform Mahathir's personality and politics (Khoo, 1995).

31 In *The Challenge*, Mahathir (1986) considers the West to have been progressively enfeebled by forgetting the very values which it once taught its colonies.

32 A major international currency realignment arranged by the leading industrial powers in September 1985 resulted in a virtual doubling of Malaysia's yen-denominated foreign borrowings with the result that, for the first time since

independence, the country experienced negative growth in that year (Jomo, 1995: 5).

33 In mid-1986, on a visit to Australia, Mahathir announced that NEP would be 'held in abeyance' (Khoo, 1995: 140). The announcement was made to the Australian Broadcasting Corporation and was publicised domestically only by the Chinese press (Jomo, 1989).

34 Bowie describes the Heavy Industries Policy as a strategy to overcome non-Malay business opposition to ethnic redistribution as well as to nurture a corps of Malay industrial managers.

35 Khoo (1995: 143) asks, 'might one not find behind the pragmatic justification of Mahathir's "abeyance" something of a logical culmination of his personal convictions and capitalism?'

36 This was reconciled with demands for state interventionism in terms of a premise of 'fair' competition. Malaysia, conversely, sees 'competition between racial groups in which one group has an absolute advantage over the other' (Mahathir, 1970: 52).

37 Jomo K. S. is of the opinion that the privatisation policy was 'certainly encouraged by the changed ideological climate of the eighties, especially with the advent of governments of the new right in the Anglo-American world' (Jomo, 1995: 6).

38 Ironically, this intense nationalism was, in part, demonstrated in opposition to Thatcherite Britain. In the early 1980s, Malaysia adopted a 'Buy British Last' policy in response to the British media's and the London Stock Exchange's hostile reception to the Malaysian government's take-over of Guthrie, the British plantation company (Khoo, 1995).

39 The so-called 'Team B' comprised Musa Hitam, the former Deputy Prime Minister and Razaleigh Hamzah who had contested for deputy premiership on no less than four previous occasions.

40 In addition to Mahathir, 'Team A' included Ghafar Baba, Deputy Prime Minister and an UMNO Vice-President.

41 The speech, made on 28 February 1991 was subsequently published in a volume on Vision 2020. All references to 'The Way Forward' are from this 1993 publication.

42 'Developed status', the ultimate goal of Vision 2020 is to be achieved by annual GDP growth of 7 per cent; taking into consideration population growth, this would mean that, 'by the year 2020, Malaysians will be four times richer (in real terms) than they were in 1990' (Mahathir, 1993: 408).

43 In addition to architecture and building design, this may be understood in terms of new 'leisurescapes' for middle-class tourists. Cartier's (1998) work on Melaka considers both the conservation of built heritage and 'ersatz' environments such as Mini-Malaysia, where visitors can view miniature representations of the country's historic house forms.

44 Shamsul suggests this may have been given extra significance in the context in which 'The Way Forward' was made: 'The concern to avoid Malaysia suffering the same fate as Yugoslavia was also in his mind when he introduced his Vision 2020 in 1991' (Shamsul (1996b: 331).

45 In *The Borderless World*, for example, Ohmae describes how 'the traditional way of looking at the trade statistics, based on a nation's balance with another nation, has become obsolete ...' (Ohmae, 1992: xi).

46 Themes of unity, neighbourliness and tolerance featured prominently in subsequent speeches on MSC, including those given during a two-month trip to the USA, Japan and Europe. It is worth noting that this 'odyssey to drum up support' (Kaur, 1997a: 1) for the MSC received far greater coverage in the Malaysian media than in that of the countries Mahathir actually visited.

Mahathir's tour points to the importance of MSC in transforming the way in which Malaysians envision themselves and appropriate national futures as much as how Malaysia is presented internationally.

4 Kuala Lumpur City Centre (KLCC)

1 In fact, even writers in the 1930s described the city as an external construction, characterised by European government buildings, banks and businesses (Emerson, 1937). For Lim Heng Kow this is evidence that Kuala Lumpur and, indeed, the entire 'urban system' of Malay(si)a was a product of colonial (British) and 'immigrant' (that is, Indian and especially Chinese) activities (Lim, 1978).

2 An urbanised zone extending south-westwards including the industrial estates of the new town of Petaling Jaya and Old Klang Road, and forming a corridor towards Port Swettenham (now Port Klang) was identified as early as the 1950s (Murphey, 1957; Hamzah, 1965). This was officially recognised as an extended Klang Valley urban region in 1972.

3 Kuala Lumpur City Centre (Holdings) Sdn Bhd were the developers and property managers for the site. I have abbreviated the company name to 'KLCC Holdings' for all references to the company in this chapter.

4 A political unit comprising the states of Perak, Selangor, Negri Sembilan and Pahang.

5 He used the English phrase (Mahathir, 1992: np).

6 The conception of transformation from 'Third World to First' was (and remains) extremely powerful in the political discourse of regional elites (see, for example, Lee, 2000).

7 Precisely when the first 'high-rise' was built, of course, depends on definition. The Malaysian architect, Ken Yeang, notes that the eight-storey Federal House was constructed as early as 1954 (Yeang, 1992). The same author considered that by the mid-1980s the Kuala Lumpur skyline resembled that of Houston, Texas (Yeang, 1987).

8 The reflection of the buildings in this representation is perhaps worthy of note. Initial designs for KLCC attempted to incorporate a lake in which the towers could be reflected. Interview with Noraishah Hussain, Project Manager, Park Development Department, KLCC Bhd, 20 May 1997.

9 This is perhaps best exemplified by descriptions of buildings' technological advancement or intelligence. KLCC is described as having 'cutting edge communications' and 'sophisticated visual, sound and data links to the rest of the world' (KLCC Holdings, 1996a: 7).

10 Global expertise refers not only to a small number of companies with experience of working on megaprojects – the Lehrher McGovern-Bovis project management team which was employed by KLCC, for example, had earlier worked on the Broadgate and Canary Wharf projects in London – but right down to the level of individuals. One informant noted that many of the experts working on the KLCC project were the 'same faces' as he had seen on previous projects in Hong Kong and Singapore. Interview with Project Manager at Zublin J. V., KLCC car and truck rental, 16 April 1997.

11 Both were designed by the Malay(sian) architect Hijjas Kasturi (see A. Tan, 1996).

12 Another record-breaking attempt was reportedly underway in China as early as 1994 (*Business Week*, 1994). Many were subsequently reported in a variety of Asian countries (see *Progressive Architecture*, 1995).

13 In his work on the competition for visibility on the skyline of Melbourne, Australia, Kim Dovey noted how 'symbolic capital is not so much created as moved around from one temporary landmark to another' (Dovey, 1992: 186).

14 In addition to Cesar Pelli and Associates as architects, the masterplan for KLCC was drawn up by Klages Carter Vail and Partners (Cesar Pelli and Associates, 1994).

15 King cites Microsoft's 3-storey Seattle headquarters as an example of the new model of corporate power – 'suburban greenery' (King, 1996: 100). Perhaps, as Jacobs points out, 'the lean glass-walled skyscraper is no longer capable of symbolically carrying capitalism' in its new 'flexible' forms (Jacobs, 1994: 356). The term 'obsolescent modernities' was used by King in a paper presented at the *Second International Symposium on Cultural Criticism* at The Chinese University of Hong Kong in 1995. Interestingly, this phrase was not used in the later published version of this conference paper (King, 1996).

16 As *Progressive Architecture* put it, 'a shift of historic proportions is taking place, and architecture is the present symbol of that transformation' (*Progressive Architecture*, 1995: 44).

17 As opposed, perhaps, to giant 'crutches' to support dependent Malays (Mahathir, cited in Jayasankaran, 1988: 9).

18 As one article on the anniversary of national independence put it: '... after 39 months of sheer hard work, ingenuity and determination, the country's latest monument to its "Malaysia Boleh" spirit is a reality that can truly stand tall among the world's greatest edifices' (Chin, 1999: 3).

19 Attention has been given to these implications in relation to another project designed by Cesar Pelli, Canary Wharf in London (see Bird, 1993).

20 In an edition of *Architecture Malaysia* devoted to debate over an appropriate future for KLCC, for example, GTA was described as 'an architectural phenomenon that has been influenced purely by commercial and not environmental objectives' (Sobri, 1985: 21).

21 Interview with President, Environmental Protection Society of Malaysia, 22 May 1997.

22 Interview with General Manager, Real Estate Management Division, Kuala Lumpur City Centre Holdings Sdn Bhd, 7 April 1997.

23 Not only the height of the building, but also its fibre-optic link and advanced air-conditioning meant that rental prices were 50 per cent higher than any other building in Kuala Lumpur in 1997. Interview with Manager, Jones Lang Wootton (JLW), 29 May 1997. JLW were the leasing agents for tower 2 and this informant conceded that there had been a 'slow uptake' of the half-million square feet of space. However, he also considered that the rental prices were very competitive in relation to other Asian cities. KLCC was not, in any case, competing with other buildings in the city – the Petronas Towers was said to be 'out on its own, in its own league'. Similarly, the General Manager of KLCC Holdings' Real Estate Management Division considered that the building was 'not in a class competing with everyone else' (Interview 7 April 1997).

24 This idea was actually based on a cartoon about the construction of the Prangin Shopping Mall in Penang which was alleged to be undermining the foundations of Komtar, the tallest building on the island. The cartoon contrasted 'now: Komtar – the tallest building in Penang' with 'in future: Komtar – the longest building in Penang' (MOESA, 1997: 12). I have alluded already to a more general critique of the 'national obsession' superlatives – 'the biggest, the longest, and the widest' (Mustafa and Subramaniam, 1995: 5).

25 'Skypricker' is the name Charles Jencks gives to 'vertical pointing towers' (Jencks, 1980: 6).

26 As one newspaper article on the 'leaning tower episode' reported, 'Back in late 1995, a persistent rumour began making the rounds in Kuala Lumpur's eateries. It was said that one of the towers was leaning. And with the naked eye, it did look slightly askew. Fears of it toppling and causing a monumental disaster soon

turned the rumour into a news story, giving it a certain legitimacy' (Murugasu, 1999: 3). A potential 'public relations nightmare' was averted when 'independent foreign consultants' concluded that the 'tilt' was well within the 'tolerance level' (*ibid.*)

27 This was ironic given that the original justification for moving the racecourse was said to be the traffic congestion in Jalan Ampang on race days (Padman and Lim, 1989)

28 A total of 83,500 square metres of stainless steel extrusions were used to clad the walls of the building (Chin, 1999).

29 In fact, this is because the floor-plan is simply not suitable for letting in these parts of the building which points to the expense associated with building so tall.

30 In my case, studying the building may certainly have heightened a perception of it looking back. But certainly in the mid-1990s, the building loomed larger and larger in lives in the city.

31 I am not implying here that the Petronas Towers is being mobilised as a panoptic technology. My concern is rather with the way in which these popular meanings given to the building contest the exercise of state power more generally. The increased surveillance capacity of contemporary cities has, however, featured prominently in recent research in urban geography and related fields (see Fyfe and Bannister, 1998; Lyon, 1994).

32 Interview with Project Manager, Zublin J. V., 16 April 1997.

33 Interview with General Manager, Real Estate Management Division, Kuala Lumpur City Centre Holdings Sdn. Bhd., 7 April 1997.

34 As in other projects, much of the manual labour was carried out by immigrant labour from Indonesia and Bangladesh. At one stage, there were some 6,000 foreign workers employed at KLCC. Interview with Head of Logistics, Lehrher McGovern, 10 May 1997.

35 Interview with Project Manager, Zublin J. V., 16 April 1997.

36 Interview with Head of Logistics, Lehrher McGovern, 10 May 1997.

37 Interview with Head of Logistics, Lehrher McGovern, 10 May 1997. Genting Highlands is the location of Malaysia's one casino, owned by a Chinese tycoon. Gambling is, of course, prohibited for Muslims. The *imam* was perhaps, therefore, completing a process of spiritual and/or Islamic purification of the Federal Territory. V. S. Naipaul noted that, on his first visit to Kuala Lumpur in the late 1970s, the racecourse had been the only place in Kuala Lumpur where gambling was permitted: 'The government was aggressively Malay and Muslim. Gambling was un-Islamic, and this week-end racecourse excitement was only a humane concession to the Chinese. They were the great gamblers' (Naipaul, 1998: 386).

5 Putrajaya and Cyberjaya

1 Relocation had earlier been planned for Janda Baik in the state of Pahang (Faezah, 1999). See Brookfield *et al.*, (1991) on earlier plans for the development in Janda Baik.

2 Indeed, it is likely that evaluation of the information technology requirements for the new administrative centre helped foment the MSC concept. Jaafar's (1995) article in *Computimes* contextualised plans for the administrative centre in terms of a 'future corridor' including the Kuala Lumpur International Airport.

3 'Route 2020' was the name given to the proposed 'Kuala Lumpur–Putrajaya–KLIA Dedicated Highway' (see, for example, Multimedia Development Corporation, 1997b: 14). The 42 km long route was intended as a 'high-speed linkage' enabling the journey from Kuala Lumpur to the new airport to

be made in 30–45 minutes. Government investment in the project was estimated at RM 1,800 million (Environmental Asia Sdn Bhd, 1997). I term this an 'imagined transect' partly in relation to the metaphor of development 'in line with Vision 2020', but partly also because the route did not exist on the ground during fieldwork in 1997.

4 From presentation in the Putrajaya stand at Quality Urban Life 97 at Putra World Trade Centre, Kuala Lumpur, 3–6 July 1997.

5 Dr Mohamed Arif Nun was a very important figure in the development of MSC. MIMOS, along with Malaysian Modernisation and Management Planning Unit (MAMPU), was responsible for the IT aspects of Putrajaya. He was Deputy Director-General of MIMOS from March 1991 to October 1996 prior to which he had lectured in Electrical Engineering at Universiti Teknologi Malaysia (UTM) (Teoh, 1997).

6 Anuar was Deputy Director-General (Information Technology) of MAMPU.

7 Mohamed Arif Nun (1996) referred to Malaysia's poor performance in the 1995 World Competitiveness Report (apparently generated by the Swiss-based World Economic Forum and the Institute of Management Development).

8 The corridor was to be founded upon a 2.5–10-gigabit digital optical fibre backbone enabling direct high-capacity links to Japan, USA and Europe (Multimedia Development Corporation, 1996a).

9 It was perhaps ironic, therefore, that the Multimedia Development Corporation was headed by Dr Othman Yeop Abdullah – Malaysia's 'Cyber Czar' – a 'technocrat' whose previous job had been Secretary General of the Primary Industries Ministry (Chin and Chee, 1999: 1).

10 Mahathir was keen in marketing to emphasise that 'we are not just developing MSC for ourselves' (cited in Kaur, 1997a: 1). Rather, MSC was Malaysia's 'gift to the world' (Mahathir, 1997b).

11 In addition to the Multimedia University, companies were encouraged to 'draw on the expertise of existing universities located within and near the corridor' such as Universiti Putra Malaysia (UPM) and Universiti Kebangsaan Malaysia (UKM) (Multimedia Development Corporation, 1996b: 24).

12 Manufacturing activities were not encouraged in the corridor as 'there is no value-added and any country can engage in these activities so we will not have any competitive advantage' (Mohamed Arif Nun, cited in *Investor's Digest*, 1997: 23).

13 The third criterion for companies seeking MSC status was the transfer of technology and/or knowledge to Malaysia.

14 Smart Schools were to dispense with the traditional rote learning methods said to have dominated the Malaysian curriculum since colonial times, replacing them with more 'self-directed' development. Teachers would come to function less as 'purveyors of knowledge' and more as 'facilitators of learning' (Multimedia Development Corporation, 1996b: 16).

15 MDC rejected my request to reproduce this image.

16 This is hardly surprising given the British colonial origins of planning in what is today Malaysia and the post-independence training of Malaysian planners in Britain and Australia. The enactment of the Town and Country Planning Act in Malaysia in 1976 further aligned the Malaysian urban planning system with practice in England (Goh, 1991; see also Ghani, 2000). Goh Ban Lee's critical review of urban planning in Malaysia in the early 1990s had also suggested the importance of severing an apparent 'umbilical cord to Britain' while not losing sight of planning's historical role as an instrument of social amelioration extending beyond land use zoning and urban design (Goh, 1991: 195).

17 The gendered language here – and when these terms are used in subsequent parts of this section – is from the original.

18 Interview with Professor Zainuddin bin Muhammad, Director-General, Federal Department of Town and Country Planning, 12 December 2000. Information from this interview is used in subsequent paragraphs in this section.

19 A 600 hectare lake was formed by damming four branches of the Sungai Chuah and Sungai Bisa.

20 This was perhaps unsurprising given that a large proportion of the population of Cyberjaya was expected to be Western and Japanese expatriates, while Muslim Malay civil servants formed the bulk of Putrajaya's resident population.

21 Interview with officer, Special Projects Unit, Department of Town and Country Planning, 28 August 1997.

22 Indeed, according to Ismail Ngah, there will be no levelling whatsoever at Cyberjaya. At Putrajaya, however, some has been necessary in order to accommodate the ceremonial architecture. Interview with Ismail Ngah, 24 September 1997.

23 There is also a point to be made here about the cultural politics of 'authentic' Malaysian landscape identity. The *kampung* is described in the Cyberjaya planning policy as 'the original vernacular landscape' (Department of Town and Country Planning, 1997: 46).

24 Lim's association with Universiti Sains Malaysia (USM) is worthy of note. He was one of a number of researchers at USM in the 1980s who drew upon and contributed to broader issues of 'appropriate technology' and 'ecodevelopment' (see Abel, 1982).

25 Shamsul has shown how *kampung* may be defined socially as well as in terms of an administrative unit with distinct physical boundaries.

26 Ministry of Housing and Local Government included the federal Department of Town and Country Planning.

27 The fact that Lim used the word 'Malaysia's' suggests a model which was not exclusively 'Malay'. Lim presented the Malay house as a source 'for Malaysian architectural identity' (Lim, 1987: 143).

28 In the Bill of Guarantees (see Multimedia Development Corporation, 1996c).

29 See, for example, the profile of Alex Kong, Chief Executive Officer of Asia Travel Network Sdn Bhd, Malaysia's first real-time Internet-based travel reservation system (MSC.Comm, 1999b).

30 The difficulties here were highlighted by the Director of Marketing of an existing 'world class' MSC status company. Following the expiry of the company's lease in Kuala Lumpur in 1998, 'half the people would have left if we moved to Cyberjaya'. Interview with Director, Marketing and Business Development, Asia South and Malaysia, Unisys MSC Sdn Bhd, 30 June 1999.

31 The 'Caring Society' concept, mentioned in 'The Way Forward', had earlier fomented debate on how Malaysia should 'manage the industrialisation process to minimise its most negative consequences and to preserve social cohesion, human dignity and quality of life' (Cho and Ismail, 1992: v).

32 'Firewall' is the term given to a corporate computer security system to bar unauthorised access.

33 On Petaling Jaya, see McGee and McTaggart (1967).

34 This is defined not in relation to some notion of 'authenticity', but rather in terms of the organic elements of the environment. For geographical work on the social construction of 'nature' in neighbouring urban Singapore, see Kong and Yeoh (1996).

35 Interview with Officer, Master Planning Department, Kuala Lumpur City Hall, 4 June 1997.

36 Ironically, the impact was somewhat diminished by the death of Princess Diana. Many Malaysians were busy following events surrounding a car crash involving

a member of the royal family of the former colonial power on 'Western' television networks.

37 In addition to Petronas, which controlled a forty per cent stake in Putrajaya Holdings, the other shareholders were Khazanah Nasional Bhd (40 per cent) and Kumpulan Wang Amanah Negara (20 per cent). The former is an investment arm of the federal government, wholly owned by the Ministry of Finance, while the latter is a government trustee body (HG Asia, 1997b).

38 Hong Leong Properties, Malaysian Plantations, Malaysian Resources Corporation, Peremba and SP Setia. Malaysian Plantations is now known as Kamunting Corporation while Malaysian Resources Corporation is no longer a joint venture partner of Putrajaya Holdings (personal communication with Eddie Khalil bin Hasbullah, Corporate Communications Department, Putrajaya Holdings Sdn Bhd, 3 July 2003).

39 This included Renong Bhd, Landmarks Bhd, Country Heights Holdings Bhd and Golden Hope Plantations. Like Putrajaya Holdings, the Cyberview Consortium also included some government interests: the Permodalan Nasional Bhd (PNB, a government trust agency incorporated to advance the *bumiputera* share of corporate equity by purchasing and holding shares on behalf of the community) and the Selangor State Government as well as MDC. As distinct from Putrajaya, however there was also foreign corporate representation in the form of Nippon Telegraph & Telephone Corporation.

40 Interview with Landscape Architect, Aspinwall Clouston Sdn Bhd, 21 July 1997.

41 Ironic given that in the MSC, as for KLCC in the previous chapter, much of the physical work of construction was carried out by foreign labour.

42 Interview with Ismail Ngah, Deputy-director of National Landscape Department, 30 May 1997. The original Landscape Unit of the Department of Town and Country planning was upgraded to a Division of Public Parks and Landscape before the creation of a separate department.

43 Interview with Ismail Ngah, 24 September 1997.

44 Interview with Ismail Ngah, 24 September 1997.

45 As this magazine of the Kuala Lumpur Stock Exchange put it, 'once successful, MSC will be cloned to other parts of the country such as Penang, Selangor, Johor, Sabah and Sarawak' (Harun, 1997: 1).

46 This is a familiar theme in utopian accounts of the social and political possibilities of 'telecommunities' (see Mulgan, 1989).

47 In response to a question as to whether MDC was 'creating a country within a country with MSC' (BBC2, 1998), its Executive Chairman replied:

> It's not the idea at all to almost create a unique territorial domain for the development of multimedia within the MSC. It is at the initial stage that the MSC will appear to be an enclave where you provide all the incentives to all the companies coming in that fall within the criteria to be given MSC status; to invest and be given the incentives. ... As the MSC develops then the development will be rolled out to the other parts of the country.
> (Dr Othman Yeop Abdullah, interviewed in BBC2, 1997)

MDC saw Cyberjaya, in particular, as a model to be 'rolled out', envisioning that, by 2020 there would be '12 intelligent cities linked to the global information highway' (Multimedia Development Corporation, 1997b: 10). In addition, in the week following the Teleconferencing Dialogue, Mahathir unveiled a National Information Technology Agenda (NITA), noting a belated recognition of MSC's 'limited scope in terms of geographical coverage and involvement of the nation' (cited in Yong and Choong, 1997: 1).

48 Quality Urban Living '97, 'the largest international forum on quality urban living' was held at the Putra World Trade Centre from 3–6 July. Multimedia Asia

1997 conference and exhibition on the theme 'moving MSC from vision to reality: benchmarking new solutions' ran from 16–19 September at the Mines Resort, Sungai Besi.

49 From Putrajaya stand at Quality Urban Life 97 at Putra World Trade Centre, Kuala Lumpur, 3–6 July 1997.

6 Beneath the intelligent cities

1 As we noted in the previous chapter, the Teleconferencing Dialogue was followed by messages in the press about how the government was responding to the geographical limitations of MSC through the National Information Technology Agenda (NITA). Other states announced plans for 'mini MSCs'. Yet, as Ho Chin Soon – whose property information company compiled the South Klang Valley Master Plan – suggested, only the Kuala Lumpur urban region had the necessary existing urban fabric (Ho, 1997).

2 I will explain the emergence of these groups and naming practices in the relevant section of the chapter.

3 The 'Gotham' spectre was raised by Deputy Prime Minister Anwar Ibrahim in his address to the conference on 3rd July (Anwar, 1997).

4 The dual meaning of the 'natural' which this suggests runs through this chapter and, as such, demands further explanation. On the one hand, 'nature' is taken to mean non-human elements and phenomena in the material world. On the other hand, nature can be used normatively, as a standard of evaluation. In so far as critique of the existing nature of Malaysian development focuses on its supposedly negative effects on the natural environment, the two meanings may be said to overlap. Environmental arguments are not, however, the only ones used against the appropriateness of MSC development either in Figure 6.1 or indeed the rest of this chapter.

5 I use the 'North' rather than 'West' here given that the discussion concerns the 'South-South' Mayors Conference.

6 From fieldwork notes taken at the *Quality Urban Life '97 Exhibition.*

7 Although Sumurcity was not part of MSC, it was included in Quality Urban Life's 'Vision Towns and Cities '97' section, 'a major exhibition on planned developments on the road to achieving Vision 2020'. Klages Carter Vail and Partners are 'the same people behind the KLCC project' (*Building and Investment*, 1997: 23).

8 In the mid-1990s, Daim was Economic Adviser to the Government and UMNO treasurer. An apparently close confidante of Mahathir, Daim had also served as Finance Minister between 1984 and 1991 (Gomez and Jomo, 1997).

9 Interview with officer, Special Projects Unit, Department of Town and Country Planning, 28 August 1997.

10 In particular, the presentation made by Nathaniel von Einsiedel from the United Nations Urban Management Program (von Einsiedel, 1997).

11 Harper was referring, in particular, to the term 'civil society' and the difficulty of extrapolating it from Western experiences to the Malaysian situation. As one article points out, though the term *masyarakat madani* ('civil society') was itself very fashionable in Malaysia, 'giving the concept meaning seems much less popular than mouthing it' (Tan, 1997: 12).

12 The fight against the Malayan Union, for example, is now often seen as a demand for the 'accountability' of the Malay rulers to the *rakyat* ('people') (Harper, 1996).

13 These were abolished in the late 1960s allegedly because of the domination of urban areas by non-Malays who were, at that time, predisposed towards the opposition political parties.

14 The Habitat Agenda does in fact acknowledge the contribution of the private sector, if only as part of a broader 'Civil Society' (United Nations Human Settlements Programme, 1996). Sia called for more rigorous enforcement of state enforcement of planning regulations (Sia, 1997a). Yet given the state-corporate complicity alluded to above, this would involve more than merely an expanded state role.

15 The 'back door' take-over by Permodolan Nasional Bhd (PNB or National Equity Corporation) led the London Stock Exchange to amend its take-over regulations. Mahathir had apparently been frustrated by an earlier sale of Guthrie's trading subsidiary and Dunlop's plantation subsidiary to Multi-Purpose Holdings Bhd, the investment arm of the Malaysian Chinese Association (MCA) (Khoo, 1995: 5).

16 Harrison and Crosfield played a 'momentous' role in the cultivation of rubber in Malaya. They began financing rubber cultivation in the years 1889–1901 although their first 'major and overt undertaking' was the Petaling Estate in Selangor planted in 1903 (Coates, 1987: 116).

17 The three villages were Kg. Sri Meranti, Kg. Limau Manis and Kg. Datuk Abu Bakar Baginda. The four plantation estates were Perang Besar, Sedgeley, Madingley and Galloway (*Universiti Pertanian Malaysia*, 1995). The subsequent institutional renaming of *Universiti Pertanian* ('agricultural') to *Universiti Putra Malaysia* was itself a toponymic indication of broader land use transformation in the southern corridor.

18 The scare-marked 'Indian' here denotes the problematic nature of this categorical singularity. While it is possible to carry out an historical deconstruction of each of Malaysia's political communities, the divisions are perhaps most marked among Indians. Apart from broad distinctions made in terms of geographical origin ('north' or 'south' Indian), religion and class are important dimensions of division. Tamil plantation workers are not only Hindus in a Malay-Muslim dominated state; they are also spatially divided from urban-based professional and public sector employed south Indians.

19 A report on the 'plight' of plantation workers in the 1970s suggested that the implementation of the 'Special Rights Program for the Malay ethnic group soon after Independence was attained in August 1957' marked the beginning of 'the systematic deprivation of the Indian community' (Gunawan and Raghavan, 1977: 5).

20 The Chinese and Malay gains were from 25 per cent to 38 per cent and 2–3 per cent to 18.2 per cent respectively (Jayasankaran, 1995b).

21 The Klang Valley South Master Plan compiled by Ho Chin Soon Research Sdn Bhd in January 1993 devoted a cartographic shade to land owned by Golden Hope. The legend read simply, 'The plantation company that has the most land banks in Klang Valley south. Congratulations' (Ho, 1997). In the Putrajaya Master Plan areas, in addition to Perang Besar estate, Golden Hope owned the 1,272 hectare Galloway Estate.

22 Community Development Centre (CDC) figures. I am grateful to S. Arutchelvan for these.

23 Ironically, in the 1920s, Perang Besar was a test-bed for new technologies and the use of new materials in commercially based research programmes. Floated in 1926, Prang Besar Estate Ltd was where 'work was carried out leading to the development of "proprietary" types of high yielding planting material used by some estates in the later 1930s and more widely post-Second World War' (Drabble, 1991: 46).

24 According to S. Arutchelvan, the federal government and Golden Hope each considered compensation to be the other's responsibility. The government claimed that the plantation company, as the employer of estate workers, was

responsible for their welfare, while the plantation company argued that since the government had become the owner of the land and the estates, compensation was their responsibility. Interview with S. Arutchelvan, 6 August 1997.

25 In particular, the Plantation Workers Support Committee and Community Development Centre (CDC). CDC was set up by a group of students from Universiti Kebangsaan Malaysia (UKM) in the mid-1980s. It now includes members both within and outside the university and deals predominantly with labour, resettlement and compensation issues. Interview with A. Sivarajan, CDC volunteer, 13 September 1997. I was able to visit Perang Besar through CDC. I am grateful to A. Sivarajan, Simon Kurunagaram and Kohila Yanasekaran for assistance with interviews at Perang Besar estate. All interviews were carried out in Tamil by CDC volunteers who translated back to me in English. All subsequent quotations from interviews in the estate are from my fieldwork notes and are not direct translations of residents' responses. I have not used the real names of estate interviewees.

26 Interview with A. Sivarajan, 13 September 1997.

27 Interview with Komathi, Perang Besar resident, 13 September 1997.

28 Interview with Nithya, Perang Besar resident, 13 September 1997.

29 Interview with Komathi, Perang Besar resident, 13 September 1997.

30 Interview with Nithya, Perang Besar resident, 13 September 1997.

31 The temple reportedly played a very important role in community organisation. One of the most frequently-cited fears of resettlement was a perceived break-down in community identity which would result from the four estates sharing a temple at the new site. The 4,400 square feet which had reportedly been allocated for this shared temple amounted to a smaller area than the Perang Besar temple alone. Interview with M. Sadayan, Perang Besar MIC Deputy Chairman and NUPW leader, 13 September 1997.

32 The suffering of Tamil plantation workers – from overwork and beatings to malnutrition and malaria – are described in Ramasamy (1994).

33 Interview with Nithya, Perang Besar resident, 13 September 1997.

34 Interview with Nasir Mohd. Shaari, Special Projects Officer, Federal Department of Town and Country Planning, 28 August 1997.

35 Both are located north of Kuala Lumpur (refer to Figure 5.1).

36 Interview with M. Sadayan, Perang Besar MIC Deputy Chairman and NUPW leader, 13 September 1997. Other residents were concerned at the prospect of sharing with the other plantation communities. Although there had been interaction among them, each considered itself as a separate community. As alluded to above, the temple was the chief binding force for the estate community at Perang Besar, but the plan at Dengkil was to construct one temple for all four communities.

37 Personal communication with Jebasingam Issace John, Director, City Planning Department, Perbadanan Putrajaya, 7 November 2000.

38 Interview with S. Arulchelvan, 6 August 1997.

39 A city in the state of Johor, at the southern end of Peninsula Malaysia (shown in Figure 3.1).

40 Nicholas notes how what were previously nineteen official sub-groups were reduced to 18 'for convenience of tabulation' (Nicholas, 2002: 121).

41 Indeed, as Nicholas shows, 'Orang Asli' is one of a succession of externally-imposed terms which construct the various indigenous minority groups of Peninsula Malaysia as a homongeneous grouping (Nicholas, 2002).

42 In East Malaysia, there is a separate term, *pribumi*.

43 *Masuk Melayu* (to 'become Malay') conventionally means to become a Muslim. While there are Orang Asli who have converted to Islam (and even a Muslim Orang Asli Welfare Association – see Nicholas, 2002) others see their 'Orang

Asli' identity – and distinction from 'Malayness' – as precisely a matter of *not* being Muslim.

44 Interview with Majid Suhut, President of *Persatuan Orang Asli Semenanjung Malaysia* (POASM, 'Orang Asli Association of Peninsula Malaysia') 9 July 1997. The remainder of this paragraph consists of material from the interview with Majid Suhut.

45 The very existence of other organisations, however, problematises any notion of a holistic or unified Orang Asli political identity.

46 Interview with Colin Nicholas, Coordinator, Centre for Orang Asli Concerns (COAC), 8 June 2002.

47 Interview with Colin Nicholas, 8 June, 2002.

48 Even prior to colonial contact, groups now labelled 'Orang Asli' were collectors of jungle produce which was highly-valued in the long-distance trade of the time (Zahid, 1990). See Short *et al.* (2000) for a recent addition to the literature on earlier rounds of globalisation.

49 The 'change of income sources is basically due to the depleting forest areas around their settlement ...' (Engineering and Environmental Consultants Sdn Bhd *et al.*, 1993: E3–35).

50 This too was, in fact, a result of external historical forces. While one historical profile traced settlement in the KLIA site area to the 1810s (Ramli, 1991), it was during the Emergency that the Temuan were moved into resettlement camps along the Sungei Salak (Tan, 1993). Nonetheless, Kampung Busut would have been evaluated positively in dominant systems of evaluation precisely because it is one of the oldest Orang Asli settlements. The fact that the eventual defeat of the Communists is often attributed to the Briggs Plan in which Chinese squatters and Orang Asli were resettled in fortified New Villages and camps to cut off supplies of food and intelligence to the guerrillas, may help to explain an historical state preference for Orang Asli settlement as opposed to dispersal.

51 As such, any opposition to relocation is cast as anti-development (see Zawawi, 1996b).

52 I am grateful to Yusof bin Alip and his 4-wheel drive for enabling me to visit the village.

53 Interview with Encik Daud Tahir, K.L. International Airport Sdn Bhd Administrative Division, 16 September 1997. According to Daud, around 8 hectares of the total resettlement area was allocated for the construction of housing and other facilities, while the remainder was intended for oil palm cultivation.

54 Interview with Encik Daud Tahir, K.L. International Airport Sdn Bhd Administrative Division, 16 September 1997.

55 Interview with Batin Senin Anak Awi, Kampung Busut Baru, Bukit Cheeding, 12 October 1997. 'Batin' is a traditional Orang Asli elder, in this case, head of the village.

56 Interview with Batin Senin anak Awi, 12 October 1997. The remainder of this paragraph is based on material from this interview.

57 This assessment was corroborated by the post-relocation decision to turn surrounding land into a Wetland Sanctuary (see below).

58 For an anthropological perspective on the spiritual and psychological value of ancestral land for the Temuan, see Zawawi (1996a). According to Wan Zawawi, resettlement extends beyond economic or material dispossession and 'strikes deep into the soul of the Orang Asli' (*ibid.* 571). Lye (2002) provides an excellent recent treatment of the environmental relationships of another Orang Asli group, the Bateks. However, it is also important to note that environmental relations generalisable to specific 'cultural' groups (let alone to Orang Asli as a whole) are increasingly problematic given a diversity of social and economic

experiences and, perhaps more significantly, the cultural politics of Orang Asli identity (see Nicholas, 2002).

59 Interview with Encik Daud Tahir, K.L. International Airport Sdn Bhd Administrative Division, 16 September 1997.

60 Interview with Batin Senin Anak Awi, 12 October 1997.

61 The Environmental Impact Assessment had noted that, at the KLIA site, the Kampung Busut community had 'constructed modern houses for themselves and are enjoying the labour of their hard work' (Engineering and Environmental Consultants Sdn Bhd *et al.*, 1993: D2–20).

62 On the problematic dichotomy of tradition–modernity and singular developmental route for Orang Asli, see Lye (2002).

63 The very fact that the swampy area was considered suitable for a wetland sanctuary appeared to confirm its unsuitability for agriculture.

64 This of course evokes Figure 6.1 which is thus understood as working to the exclusion of the Orang Asli too; dispossessed to make way for an Edenic Urban Utopia and, lying outside its sealed environment, vulnerable to Gotham's negative externalities.

7 Conclusion

1 See also Blaut (1993).

2 The very 'global' economic forces which had promised a 'fully developed Malaysia' and/or 'Multimedia Utopia' were now cast by Mahathir as at best unfair, but more likely 'criminal' (cited in Ashraf, 1997: 1). This apparent rejection of economic liberalisation was perhaps less of a turn-around than imagined at the time by international media and fund managers. Mahathir's outbursts against 'rogue' currency 'manipulators' – most famously George Soros – who 'are no longer playing the normal way and are acting like a monopoly' (cited in *The Star*, 1997b: 2) and so preventing the free and fair operation of the market, had resonances with arguments in *The Malay Dilemma* which promoted state protection of Malays in the face of 'Chinese' economic domination. Mahathir's suggestions that speculative currency attacks could be the work of Jews – 'we are Moslems and the Jews are not happy to see the Moslem progress' (cited in Reuters, 1997a) – was also hardly novel or even surprising in the context of Malay(sian) politics. The point was that comments made to a crowd of 10,000 in predominantly Malay Terengganu – and which provoked strong criticism from the US State Department (Reuters, 1997b) – were not intended for international consumption.

3 This began in September 1998 as Gerak (*Majlis Gerakan Keadilan Nasio*nal or 'Council of the National Justice Movement'), a coalition of PAS, DAP, Parti Rakyat Malaysia and NGOs including ADIL, the precursor to Parti keADILan Nasional ('National Justice Party') which was led by Anwar Ibrahim's wife, Azizah Wan Ismail (see Khoo, 2000).

4 Including one which assumed RM 1.2 billion of debt of Konsortium Perkapalan, the shipping firm of Mahathir's eldest son (Saravanamuttu, 2003).

5 Petronas increased its stake in Putrajaya Holdings from 40 to 64.4 per cent, while Khazanah now holds only 15.6 per cent (down from 40 per cent) (personal communication with Eddie Khalil bin Hasbullah, Corporate Communications Department, Putrajaya Holdings Sdn Bhd, 3 July 2003).

6 As Ziauddin Sardar put it, 'the greatest annoyance of all in a groundswell of dissent that would not simply fade away was that the Internet, the basis of Mahathir's megalomaniacal dreams, became the locus of all the information that he would not permit to circulate in real-time reality' (Sardar, 2000: 234). Attempts to curtail such information flows have been balanced against MSC

promises (Abbott, 2001). Abbot has also rightly cautioned against celebratory over-statement of the Internet's potential for democratisation and political transformation.

7 From property investment flyer, 'Cyberia: A dot-com property?'. An article in *Asiaweek* also noted the comparison with 'Siberia', although by 2001 many of the Cyberia properties had in fact been sold (Mellor, 2001).

8 Applications for MSC status 'surged' following the lifting of controls (Indergaard, 2003); work on transport infrastructure in KLMA has resumed, including the completion of the Express Rail Link (ERL) which connects Kuala Lumpur to KLIA via Putrajaya; there are reports of emergent intelligent 'local heroes' and 'world class Malaysian IT companies' (*Asiaweek.com*, 2001); and there have been further proactive responses to an apparent shortage of knowledge workers within and beyond MSC.

9 Landmark cases from Canada, Australia and Nigeria were drawn upon in proprietary Temuan 'native' claims for compensation for their displacement. The fate of Sagong bin Tasi and the other plaintiffs at Bukit Tampoi were thus tied to the distant places of other 'aborigines' and 'indigenous people': Mabo and the Wik Peoples in Queensland, the Miriuwung and Gajerrong in Western Australia, Amodu Tijani in Southern Nigeria and the Delgamuukw in British Columbia.

10 His articles originally appeared as a column on malaysiakini.com. They have since been compiled (Farish, 2002).

References

Aban, A. (2001) 'Tales of Petronas, the world's tallest towers', *Bulatlat.com*. Online. Available HTTP: http://www.bulalat.com/news/034kl.html (11 October 2001).

Abdul Aziz Abdul Rahman and Pillai, S. (1996) *Mahathir: Leadership and Vision in Science and Technology*, Serdang: Universiti Pertanian Malaysia Press.

Abel, C. (1982) 'Living in a hybrid world; the evolution of cultural identities in the developing nations', *Design Studies* 3, 3: 127–32.

Abbott, J. P. (2001) 'Democracy@internet.asia? The challenges to the emancipatory potential of the net: lessons from China and Malaysia, *Third World Quarterly* 22, 1: 99–114.

Ackerman, S. and Lee, R. (1988) *Heaven in Transition: Non-Muslim Religious Innovation and Ethnic Identity in Malaysia*, Honolulu: University of Hawaii Press.

Adam, C. and Cavendish, W. (1995) 'Background', in K. S. Jomo (ed.) *Privatizing Malaysia*, Boulder: Westview Press.

Adshead, R. (1989) *Mahathir of Malaysia*, London: Hubiscus.

Alatas, S. H. (1977) *The Myth of the Lazy Native*, London: Frank Cass.

Alina Rastam (1997) 'KLIA – more than just an airport', *New Straits Times*, 11 September: 11.

Aliran Monthly (1996) 'Thinking allowed', 16, 9: 19.

Andaya, B. W. and Andaya, L. Y. (2001) *A History of Malaysia* (second edition), Houndmills: Palgrave.

Anderson, B. (1998) *The Spectre of Comparisons: Nationalism: Southeast Asia and the World*, London: Verso.

Anwar Ibrahim (1996) *The Asian Renaissance*, Singapore: Times Books.

—— (1997) Address at *South-South Mayors Conference: Developing Solutions for Cities of the 21st Century*, Putra World Trade Centre, Kuala Lumpur, 3–4 July.

Appadurai, A. (1988) 'Putting hierarchy in its place', *Cultural Anthropology* 3, 1: 36–49.

—— (1996) *Modernity at Large: Cultural Dimensions of Globalization*, Minnesota: University of Minnesota Press.

Arutchelvan, S. (2000) 'Monthly wage: Why hasn't this issue been resolved?', *Aliran Monthly* 20, 11/12: 9–10.

Ascherson, N. (1998) 'Finally I meet the man I was sent to kill 40 years ago – and he smiles', *The Observer*, 14 June: 3.

Ashraf Abdullah (1997) 'Dr M: It's George Soros', *New Sunday Times*, 27 July: 1.

Ashraf Abdullah and Radzi Sapiee (1997) 'MSC flagship applications: PM advises states to wait', *New Straits Times*, 16 April: 3.

Asian Editor (1997) 'Dr Mahathir Mohamad: architect of developed Malaysia', Feb-March: 28–46.

Asian Strategy and Leadership Institute (1997) 'Introduction', *South-South Mayors Conference: Developing Solutions for Cities of the 21st Century*, Putra World Trade Centre, Kuala Lumpur, 3–4 July.

Asiaweek.com. (2001) 'Malaysia tries to spawn its own local heroes', 17 August. Online. Available HTTP: htp://www.asiaweek.com/asiaweek/technology/article/0,8707,170517,00.html (27 August 2001).

Asohan, A. (1997) 'The way ahead: ideas, concepts and challenges for our national aspirations', *The Star*, 26 June (In-Tech section): 45.

Azizan, Z. A. (1997) 'A city in the making: A case study on Putrajaya'. Paper presented at *South-South Mayor's Conference: Developing Solutions for Cities of the 21st Century*, Putra World Trade Centre, Kuala Lumpur, 4 July.

Barnett, C. (1999) 'Culture, government and spatiality: reassessing the "Foucault effect" in cultural-policy studies', *International Journal of Cultural Studies* 2, 3: 369–97.

Barter, P. (2002) 'Transport and Housing Security in the Klang Valley, Malaysia', *Singapore Journal of Tropical Geography* 23, 3: 268–87.

Barthes, R. (1979) *The Eiffel Tower and Other Mythologies*, trans. R. Howard, New York: Hill and Wang.

BBC2 (1998) 'The Net', 8 March.

Benjamin, G. and Chou, C. (2002) 'Introduction', in G. Benjamin and C. Chou (eds) *Tribal Communities in the Malay World: Historical, Cultural and Social Perspectives*, Singapore: Institute of Southeast Asian Studies.

Berman, M. (1982) *All That is Solid Melts Into Air: The Experience of Modernity*, London: Verso.

Bird, J. (1993) 'Dystopia on the Thames', in J. Bird, B. Curtis, T. Putnam, G. Robertson and L. Tickner (eds) *Mapping the Futures: Local Cultures, Global Change*, London: Routledge.

Blaut, J. (1993) *The Colonizer's Model of the World: Geographical Diffusionism and Eurocentric History*, New York: The Guilford Press.

Boey, M. (2002) '(Trans)national realities and imaginations: the business and politics of Malaysia's Multimedia Super Corridor', in T. Bunnell, L. B. W. Drummond and K. C. Ho (eds) *Critical Reflections on Cities in Southeast Asia*, Singapore: Times Academic Press.

Bowie, A. (1991) *Crossing the Industrial Divide: State, Society and the Politics of Economic Transformation in Malaysia*, New York: Colombia University Press.

Bowie, P. (1997) 'Malaysia's towers of strength', *New Straits Times*, 28 April (Life and Times section): 8.

Brenner, N. (1998) 'Between fixity and motion: accumulation, territorial organization and the historical geography of spatial scales', *Environment and Planning D: Society and Space* 16: 459–81.

—— (1999) 'Beyond state-centrism? Space, territoriality, and geographical scale in globalization studies', *Theory and Society* 28, 1: 39–78.

Brockway, L. H. (1979) *Science and Colonial Expansion: The Role of the British Royal Botanic Gardens*, New York: Academic Press.

Brookfield, H., Abdul Samad Hadi and Zaharah Mahmud (1991) *The City in the Village: The In-situ Urbanization of Villages, Villagers and Their Land Around Kuala Lumpur, Malaysia*, Singapore: Oxford University Press.

Brown, R. (1994) *The State and Ethnic Politics in Southeast Asia*, London: Routledge.

Building and Investment (1997) 'Big returns for Sumurwang', Quality Urban Life Exhibition Special Supplement, July/August: 22–3.

Building Property Review (1995) 'Kuala Lumpur City Centre: "Towering" challenges for its builders', April–June: 56–8.

Bunnell, T. (1999) 'Views from above and below: the Petronas Twin Towers and/in contesting visions of development in contemporary Malaysia', *Singapore Journal of Tropical Geography* 201: 1–23.

—— (2002a) 'Multimedia utopia? A geographical critique of high-tech development in Malaysia', *Antipode: A Radical Journal of Geography* 34, 2: 265–95.

—— (2002b) 'Cities for nations? Examining the city–nation-state relation in Information Age Malaysia', *International Journal of Urban and Regional Research* 26, 2: 284–98.

—— (2002c) '*Kampung* rules: landscape and the contested government of urban(e) Malayness', *Urban Studies* 39, 9: 1686–701.

—— (2002d) '(Re)positioning Malaysia: High-tech networks and the multicultural rescripting of national identity', *Political Geography* 21, 1: 105–24.

—— (2003) 'From nation to networks and back again: transnationalism, class and national identity in Malaysia'. In Yeoh, B. S. A. and Willis, K. (eds) *State/Nation/Transnation: Perspectives of Transnationalism in the Asia-Pacific*. London: Routledge.

—— (2004) 'Re-viewing the *Entrapment* controversy: Megaprojection, (mis)representation and postcolonial performance', *Geojournal*.

Bunnell, T., Barter, P. and Morshidi Sirat (2002) 'Kuala Lumpur Metropolitan Area: A globalizing city-region', *Cities* 19, 5: 357–70.

Business Times (1996) 'Race to the top', 18 April: 2.

Business Week (1994) 'Talk about getting uppity', 4 April: 16.

Cairncross, F. (1997) *The Death of Distance: How the Communications Revolution Will Change our Lives*, Boston: Harvard Business School Press.

Cairns, S. (2002) 'Troubling real-estate: Reflecting on urban form in Southeast Asia', in T. Bunnell, L. B. W. Drummond and K. C. Ho (eds) *Critical Reflections on Cities in Southeast Asia*, Singapore: Times Academic Press.

Camroux, D. (1994) *Looking East … and Inwards: Internal Factors in Malaysian Foreign Relations During the Mahathir Era, 1981–1994*, Centre for Study of Australia–Asia Relations: Griffith University.

Cartier, C. L. (1998) 'Megadevelopment in Malaysia: From heritage landscapes to "leisurescapes" in Melaka's tourism sector', *Singapore Journal of Tropical Geography* 19, 2: 151–76.

Castells, M. (1989) *The Informational City: Information Technology, Economic Restructuring and the Urban-Regional Process*, Oxford: Blackwell.

—— (1996) *The Rise of the Network Society*, Oxford: Blackwell.

Castells, M. and Hall, P. (1994) *Technopoles of the World: The Making of Twenty-First Century Industrial Complexes*, London: Routledge.

Cesar Pelli and Associates (1994) *Kuala Lumpur City Centre Project Brief*.

—— (1997) '… unique, unlike any Western skyscraper', *Concepts*, March/April: 28–31.

Chan, K. E. (1994) 'Internal migration in a rapidly developing country: The case of Peninsula Malaysia', *Malaysian Journal of Tropical Geography* 25, 2: 69–78.

Chandra Muzaffar (1987) *Islamic Resurgence in Malaysia*, Petaling Jaya: Penerbit Fajar Bakti.

Chandran, P. (1994) 'The forsaken lot', *The Sun*, 5 May: 10.

Cheah, S. C. (1993) 'PAM proposes a new style of living called Kampung-minium', *The Star*, 23 February (Business section): 2.

Cheng, L. and Yang, P. Q. (1998) 'Global interaction, global inequality and migration of the highly trained to the United States', *International Migration Review* 32: 626–53.

Cherry, G. E. (1988) *Cities and Plans: The Shaping of Urban Britain in the Nineteenth and Twentieth Centuries*, London: Edward Arnold.

Chin, G. (1999) 'From vision to reality', *The Star*, 31 August (Section 2): 2–3.

Chin, J. and Chee Y. H. (1999) 'Cyber Czar', *Asia Inc*, August: 16–21.

Cho, K. S. and Ismail Muhd Salleh (1992) 'Introduction', in K. S. Cho and Ismail Muhd Salleh (eds) *Caring Society: Emerging Issues and Future Directions*, Kuala Lumpur: Institute of Strategic and International Studies.

City Hall Kuala Lumpur (2003) *Draft Structure Plan Kuala Lumpur 2003: A World Class City*, Kuala Lumpur: City Hall Kuala Lumpur.

Cleary, M. and Shaw, B. (1994) 'Ethnicity, development and the New Economic Policy: the experience of Malaysia', *Pacific Viewpoint* 35, 1: 83–106.

Clifford, J. (1997) *Routes: Travel and Translation in the Late Twentieth Century*, Cambridge MA: Harvard University Press.

Coates, A. (1987) *The Commerce in Rubber: The First 250 Years*, Singapore: Oxford University Press.

Comber, L. (1983) *13 May 1969: A Historical Survey of Sino-Malay Relations*, Singapore: Graham Brash.

Corey, K. E. (2000) 'Electronic space: Creating cyber communities in Southeast Asia', in M. I. Wilson and K. E. Corey (eds) *Information Techtonics*, Chichester: John Wiley.

Cresswell, T. (1996) *In Place/Out of Place: Geography, Ideology and Transgression*, Minneapolis: University of Minneapolis.

—— (2000) '*Falling Down*: Resistance as diagnostic', in J. Sharp, P. Routledge, C. Philo and R. Paddison (eds) *Entanglements of Power: Geographies of Domination and Resistance*, London: Routledge.

Crouch, H. (1985) *Economic Change, Social Structure and the Political System in Southeast Asia*, Singapore: Institute of Southeast Asian Studies.

—— (1992) 'Authoritarian trends, the UMNO split and the limits to state power', in J. S. Kahn and F. Loh (eds) *Fragmented Vision: Culture and Politics in Contemporary Malaysia*, Sydney: Allen and Unwin.

Crush, J. (1995) 'Introduction: imagining development', in J. Crush (ed.) *Power of Development*, London: Routledge.

Daly, M. (1994) 'The road to the twenty-first century: the myths and miracles of Asian manufacturing', in S. Corbridge, R. Martin and N. Thrift (eds) *Money, Power and Space*, Oxford: Blackwell.

Danapal, G. (1992) 'A visionary development in line with 2020 goal, says PM', *New Straits Times* 16 September: 1.

Daniels, S. (1988) 'The political iconography of woodland in later Georgian England', in D. Cosgrove and S. Daniels (eds) *The Iconography of Landscape:*

Essays on the Symbolic Representation, Design and Use of Past Environments, Cambridge: Cambridge University Press.

—— (1993) *Fields of Vision: Landscape Imagery and National Identity in England and the United States,* Cambridge: Polity Press.

Daniels, S. and Cosgrove, D. (1988) 'Iconography and Landscape', in D. Cosgrove and S. Daniels (eds) *The Iconography of Landscape,* Cambridge: Cambridge University Press.

Das, K. (1981) 'A tough guy takes over', *Far Eastern Economic Review,* 30 October: 30–1.

—— (1982) 'Mahathir's "restoration"', *Far Eastern Economic Review,* 11 June: 38–41.

Davis, M. (1992) 'Fortress Los Angeles: the militarization of urban space', in M. Sorkin (ed.) *Variations on a Theme Park: The New American City and the End of Public Space,* New York: Hill and Wang.

de Certeau, M. (1985) 'Practices of space', in M. Blonsky (ed.) *On Signs,* Oxford: Blackwell.

Dean, M. (1994) *Critical and Effective Histories: Foucault's Methods and Historical Sociology,* London: Routledge.

—— (1999) *Governmentality: Power and Rule in Modern Society,* London: Sage.

Dennis, R. (1994) 'Interpreting the apartment house: modernity and metropolitanism in Toronto, 1900–1930', *Journal of Historical Geography* 20, 3: 305–22.

Department of Town and Country Planning (1996) 'Creating the vision for Putrajaya'. Paper presented at *Shaping the Vision of a City,* 16–17 September, Hotel Syuen, Ipoh.

—— (1997) *Landscape Master Plan for Cyberjaya, Malaysia,* Kuala Lumpur: Ministry of Housing and Local Government.

—— (2001) *Total Planning and Development Guidelines* (second edition), Kuala Lumpur: Ministry of Housing and Local Government.

Dicken, P. (1998) *Global Shift: Transforming the World Economy* (third edition), London: Paul Chapman.

Dicken, P., Peck, J. and Tickell, A. (1997) 'Unpacking the global', in R. Lee and J. Wills (eds) *Geographies of Economies,* London: Arnold.

Dirlik, A. (1994) 'The postcolonial aura: Third World capitalism in the age of global capitalism', *Critical Inquiry* 20, 4: 328–56.

—— (1998) 'Introducing the Pacific', in A. Dirlik (ed.) *What's in a Rim? Critical Perspectives on the Pacific Region Idea,* Lanhan: Rowman and Littlefield.

Domosh, M. (1988) 'The symbolism of the skyscraper: Case studies of New York's first tall building', *Journal of Urban History:* 14, 3: 320–45.

—— (1989) 'A method for interpreting landscape: A case study of the New York World Building', *Area* 21, 4: 347–55.

Donald, J. (1992) 'Metropolis: the city as text', in R. Bocock and K. Thompson (eds) *Social and Cultural Forms of Modernity,* Cambridge: Polity Press.

Douglass, M. (1998) 'World city formation on the Asia Pacific Rim: poverty, "everyday" forms of civil society and environmental management', in M. Douglass and J. Friedmann (eds) *Cities For Citizens,* Wiley: Chichester.

Dovey, K. (1992) 'Corporate towers and symbolic capital', *Environment and Planning B: Planning and Design* 19, 2: 173–88.

Drabble, J. H. (1991) *Malayan Rubber: The Interwar Years,* Houndmills and London: Macmillan.

—— (2000) *An Economic History of Malaysia: The Transition to Modern Economic Growth*, London: Macmillan.

du Gay, P. (1996) 'Organizing identity: entrepreneurial governance on public management', in S. Hall and P. du Gay (eds) *Questions of Culturals Identity*, London: Sage.

Duncan, J. (1990) *The City as Text: The Politics of Landscape Interpretation in the Kandyan Kingdom*, Cambridge: Cambridge University Press.

Duncan, J. and Duncan, N. (1988) '(Re)reading the landscape', *Environment and Planning D: Society and Space* 6, 2: 117–26.

Dynes, M. (1994) 'Mahathir says credibility is at stake', *The Times*, 4 April: 2.

Edwin, J. (1990) 'Paradise in Turf Club', *New Straits Times*, 9 February (Section Two): 9.

Einhorn, B. and Prasso, S. (1999) 'High-tech folly', *Business Week* (Asian edition), 22 March: 18–22.

Emerson, R. (1937) *Malaysia: A Study in Direct and Indirect Rule*, New York: Macmillan.

Engineering and Environmental Consultants Sdn Bhd, Ranhill Bersekutu Sdn Bhd and Environmental Resources Ltd (1993) *Kuala Lumpur International Airport: Environmental Impact Assessment Study* (Final Report vol. 2, annexes A–E), Kuala Lumpur: Kementerian Pengangkutan Malaysia.

Environmental Asia Sdn Bhd (1997) *Preliminary Environmental Impact Assessment: Kuala Lumpur – Putrajaya – KLIA Dedicated Highway*, Petaling Jaya: Environmental Asia Sdn Bhd.

Escobar, A. (1995) *Encountering Development*, Princeton: Princeton University Press.

Evers, H.-D. and Korff, R. (2000) *Southeast Asian Urbanism: The Meaning and Power of Social Space*, Singapore: Institute of Southeast Asian Studies.

Eyre, J. and Dwyer, D. (1996) 'Ethnicity and industrial development in Penang, Malaysia', in D. Dwyer and D. Drakakis-Smith (eds) *Ethnicity and Development: Geographical Perspectives*, Chichester: John Wiley and Sons.

Faezah Ismail (1999) 'Local flavour for a modern undertaking', *New Sunday Times*, 4 April: 15.

Farish A. Noor (1996) 'Youth culture and Islamic intelligentsia: ignoring the popular cultural discourse', *Muslimedia International*, April. Online. Available HTTP: http://www.muslimedia.com/archives/sea/sea.htm (1 December 1997).

—— (2002) *The Other Malaysia: Writings on Malaysia's Subaltern History*, Kuala Lumpur: Silverfishbooks.

Fauza A. G. (1994) 'Structural change and regional growth in Malaysia, 1970–90', *Malaysian Journal of Tropical Geography* 25, 2: 79–89.

Fergusson, P. C. (1997) *Malaysia's Multimedia Super Corridor: Techno-Push into Utopia?*, Menlo Park, CA: SRI Consulting.

Foucault, M. (1982) 'The subject and power', in R. Dreyfus and P. Rabinow (eds) *Michel Foucault: Beyond Structuralism and Hermeneutics*, Brighton: Harvester.

—— (1988) 'Technologies of the self', in L. H. Martin, H. Gutman and P. H. Hutton (eds) *Technologies of the Self: A Seminar With Michel Foucault*, London: Tavistock.

—— (1990) *The History of Sexuality Volume 1: An Introduction*, trans. R. Hurley, New York: Vintage Books.

—— (1991) 'Governmentality', in G. Burchell, C. Gordon and P. Miller (eds) *The Foucault Effect: Studies in Governmentality*, Hemel Hempstead: Harvester Wheatsheaf.

Fukuyama, F. (1992) *The End of History and the Last Man*, London: Hamish Hamilton.

—— (1998) 'Asian values and the Asian crisis', *Commentary*, February: 23–7.

Furnivall, J. S. (1939) *The Netherlands Indies: A Study of Political Economy*, Cambridge: Cambridge University Press.

Fyfe, N. and Bannister, J. (1998) 'The eyes upon the street', in N. Fyfe (ed.) *Images of the Street: Planning, Identity and Control in Public Space*, London: Routledge.

Gates, B. (1995) *The Road Ahead*, London: Penguin.

Ghani Salleh (2000) *Urbanisation and Regional Development in Malaysia*, Kuala Lumpur: Utusan Publications.

Ghani Salleh and Lee, L. M. (1997) *Low-Cost Housing in Malaysia*, Kuala Lumpur: Utusan Publications.

Giddens, A. (1984) *The Constitution of Society*, Cambridge: Polity Press.

—— (1990) *The Consequences of Modernity*, Cambridge: Polity Press.

Goddard, J. and Richardson, R. (1996) 'Why geography will still matter: What jobs go where?', in W. H. Dutton (ed.) *Information and Communication Technologies: Visions and Realities*, Oxford: Oxford University Press.

Goh Ban Lee (1991) *Urban Planning in Malaysia: History, Assumptions and Issues*, Petaling Jaya: Tempo Publishing.

Goh Beng Lan (2002a) *Modern Dreams: An Inquiry into Power, Cultural Production and the Cityscape in Contemporary Urban Penang, Malaysia*, Ithaca, NY: Cornell Southeast Asia Program Publications.

—— (2002b) 'Rethinking modernity: State, ethnicity and class in the forging of a modern urban Malaysia', in C. J. W.-L. Wee (ed.) *Local Cultures and the 'New Asia': The State, Culture and Capitalism in Southeast Asia*, Singapore: Institute of Southeast Asian Studies.

Goh C. T. (1971) *The May Thirteenth Incident and Democracy in Malaysia*, Kuala Lumpur: Oxford University Press.

Goh, R. and Yeoh, B. S. A. (eds) (2003) *Theorizing the Southeast Asian City as Text: Urban Landscapes, Cultural Documents and Interpretative Experiences*, Singapore: World Scientific Publishing.

Gomez, E. T. (1994) *Political Business: Corporate Involvement of Malaysian Political Parties*, Cairns: Centre for Southeast Asian Studies, James Cook University.

Gomez, E. T. and Jomo, K. S. (1997) *Malaysia's Political Economy: Politics, Patronage and Profits*, Cambridge: Cambridge University Press.

—— (1999) 'Afterword: from economic to political crisis', in E. T. Gomez and K. S. Jomo, *Malaysia's Political Economy: Politics, Patronage and Politics* (second edition), Cambridge: Cambridge University Press.

Gottman, J. (1966) 'Why the skyscraper?', *Geographical Review* 56, 3: 190–212.

Graham, S. (1998) 'The end of geography or the explosion of space? Conceptualising space, place and information technology', *Progress in Human Geography* 22, 4: 165–85.

Graham, S. and Marvin, S. (1996) *Telecommunications and the City: Electronic Spaces, Urban Places*, London: Routledge.

—— (1999) 'Planning cyber-cities? Integrating telecommunications into urban planning', *Town Planning Review* 70: 89–114.

—— (2001) *Splintering Urbanism: Networked Infrastructures, Technological Mobilities and the Urban Condition*, London and New York: Routledge.

Gregory, D. (1994) *Geographical Imaginations*, Oxford: Blackwell.

Guinness, P. (1994) 'The politics of identity: Malay squatters in industrialising Malaysia', in A. Gomes (ed.) *Modernity and Identity: Asian Illustrations*, Victoria: La Trobe University Press.

Gullick, J. (1955) 'Kuala Lumpur 1880–1895', *Journal of the Malayan Branch of the Royal Asiatic Society* 26, 1: 7–172.

—— (1964) *Malaya*, London: Longmans.

Gunawan, B. and Raghavan, R. (1977) *The Plight of the Plantation Workers in West Malaysia, with Special Reference to Indian Labour*, Amsterdam: University of Amsterdam.

Gupta, A. and Ferguson, J. (1992) 'Beyond "culture": space, identity, and the politics of difference', *Cultural Anthropology* 7, 1: 6–23.

Halim Salleh (1992) 'Peasants, proletarianisation and the state; FELDA settlers in Pahang', in J. S. Kahn and F. Loh (eds) *Fragmented Vision: Culture and Politics in Contemporary Malaysia*, Sydney: Allen and Unwin.

Hall, P. (1988) *Cities of Tomorrow: An Intellectual History of Urban Planning and Design in the Twentieth Century*, Oxford: Blackwell.

Hall, S. (1992) 'The West and the rest', in S. Hall and B. Gieben (eds) *Formations of Modernity*, Cambridge: Polity Press.

—— (1997) 'The work of representation', in S. Hall (ed.) *Representation: Cultural Representations and Signifying Practices*, London: Sage.

Hamzah Sendut (1965) 'The structure of Kuala Lumpur: Malaysia's capital city', *Town Planning Review* 36, 2: 49–66.

Hannah, M. (1998) 'Space and the structuring of disciplinary power: an interpretative review', *Geografiska Annaler* 79B, 3: 171–80.

Harper, T. N. (1996) 'New Malays, new Malaysians: nationalism, society and history', *Southeast Asian Affairs 1996*, Singapore: Institute of Southeast Asian Studies.

—— (1999) *The End of Empire and the Making of Malaya*, Cambridge: Cambridge University Press.

Harun Darauh (1997) 'No pipe dream', *Investors Digest* March: 1.

Harvey, D. (1989) *The Condition of Postmodernity: An Enquiry Into the Origins of Cultural Change*, Oxford: Blackwell.

Harvey, P. (1996) *Hybrids of Modernity: Anthropology, the Nation State and the Universal Exhibition*, London: Routledge.

Hefner, R. W. (2001) Cultural pluralism in Malaysia and Indonesia: Varieties of Malay multiculturalism. Paper presented at *Third International Malaysian Studies Conference*, Universiti Kebangsaan Malaysia, 6–8 August.

Held, D. (1992) 'Liberalism, Marxism and democracy', in S. Hall, D. Held and T. McGrew (eds) *Modernity and Its Futures*, Cambridge: Polity Press.

Heng, P. K. (1992) 'The Chinese business elite of Malaysia', in R. McVey (ed.) *Southeast Asian Capitalists*, Ithaca, N.Y.: Cornell Southeast Asian Studies Program.

HG Asia (1997a) 'The making of Putrajaya', *The Edge*, 3 March: 9.

—— (1997b) *Putrajaya: Market Mover 1997*, February.

Hiebert, M. (1996) 'Dr Feelgood', *Far Eastern Economic Review*, 24 December: 18–21.

Hiebert, M., Jayasankaran, S., Miller, M. and Tiglao, R. (1997) 'Future Shock', *Far Eastern Economic Review*, 27 February: 44–6.

Hilley, J. (2001) *Malaysia: Mahathirism, Hegemony and the New Opposition*, London: Zed Books.

Hillner, J. (2000) 'Venture capitals: Mapping the 46 hot spots of the global boom', *Wired*, July: 258–71.

Hing A. Y. (2000) 'Migration and the reconfiguration of Malaysia', *Journal of Contemporary Asia* 30: 221–45.

Hirschman, C. (1986) 'The making of race in colonial Malaya: political economy and racial category', *Sociological Forum*, Spring: 330–61.

Ho, C. S. (1997) 'Seven MSC real estate points'. Paper presented at *Multimedia Super Corridor: Business, Venture Capital and Real Estate Opportunities*, Marina Mandarin, Singapore, 29 May.

Ho, J. (1991) 'A plot of gold', *Malaysian Business*, 1–15 July: 15–16.

Holston, J. (1989) *The Modernist City: An Anthropological Critique of Brasilia*, Chicago: University of Chicago Press.

Hong, C. (1997a) 'Intelligent city brain of MSC', *New Straits Times*, 17 May: 16.

—— (1997b) 'Government to be proactive in formulating cyberlaws', *New Straits Times*, 26 April: 5.

Hood Salleh (1995) 'Epilogue', in Razha Rashid (ed.) *Indigenous Minorities of Peninsula Malaysia: Selected Issues and Ethnographies*, Kuala Lumpur: Intersocietal & Scientific.

Howells, J. and Roberts, J. (2000) 'From innovation systems to knowledge systems', *Prometheus* 18: 17–31.

Hussin Mutalib (1993) *Islam in Malaysia: From Revivalism to Islamic State*, Singapore: Singapore University Press.

Hutnyk, J. (1999) 'Semifeudal cybercolonialism: Technocratic dreamtime in Malaysia', in J. Bosme (ed.) *Read Me! ASCII Culture and the Revenge of Knowledge*, New York: Autonomedia.

Huxtable, A. L. (1984) *The Tall Building Artistically Reconsidered: The Search For a Skyscraper Style*, New York: Pantheon Books.

Ibrahim Ariff and Goh, C. C. (1998) *Multimedia Super Corridor*, Kuala Lumpur: Leeds Publications.

Indergaard, M. (2003) 'The webs they weave: Malaysia's Multimedia Super-corridor and New York City's Silicon Alley', *Urban Studies* 40, 2: 379–401.

Investor's Digest (1997) 'Quantum leap into cyberspace', March: 5.

Ismawi Zen (1996) 'Vision of an Islamic city'. Paper presented at *Shaping the Vision of a City*, Hotel Syuen, Ipoh, 16–17 September.

Jaafar, M. A. (1995) 'A vision to be attained in Putrajaya', *New Straits Times*, 26 October (Computimes section): 5.

Jacobs, J. M. (1994) 'The battle of Bank Junction: the contested iconography of capital', in S. Corbridge, R. Martin and N. Thrift (eds) *Money, Power and Space*, Oxford: Blackwell.

Jackson, J. (1968) *Planters and Speculators: Chinese and European Agricultural Enterprise in Malaya, 1786–1921*, Kuala Lumpur: University of Malaya Press.

Jackson, P. (1998) 'Domesticating the street: the contested spaces of the high street and the mall', in N. R. Fyfe (ed.) *Images of the Street: Planning, Identity and Control in Public Space*, London: Routledge.

Jackson, R. (1991) *The Malayan Emergency: The Commonwealth's Wars 1948–66*, London: Routledge.

Jamilah Ariffin (1994) *From Kampung To Urban Factories: Findings From the HAWA Study*, Kuala Lumpur: University of Malaya Press.

Jayasankaran, S. (1988) 'Premier in power', *Malaysian Business*, 1 January: 5–11.

—— (1995a) 'The chosen few: privatization allows Mahathir to pick winners', *Far Eastern Economic Review*, 21 December: 30.

—— (1995b) 'Balancing act', *Far Eastern Economic Review*, 21 December: 24–6.

Jencks, C. (1980) *Skyscrapers, Skyprickers, Skycities*, New York: Rizzoli.

Jerald Gomez and Associates (2002) *Plaintiff's Written Submission, MTI-21-314-1996*, Kuala Lumpur: Jerald Gomez & Associates.

Jessop, B. and N-L. Sum (2000) 'An entrepreneurial city in action: Hong Kong's emerging strategies in and for (inter)urban competition', *Urban Studies* 37: 2287–313.

Jesudason, J. V. (1990) *Ethnicity and the Economy: The State, Chinese Business and Multinationals in Malaysia*, Singapore: Oxford University Press.

John, J. I. (2000) 'The planning and urban design of Putrajaya'. Paper presented at *Seminar on Architecture, Urban Design and Development*, APDC, Kuala Lumpur, 7 November.

Johnson, C. (1982) *MITI and the Japanese Miracle*, Stanford, CA: Stanford University Press.

Johnson, N. (1995) 'Cast in stone: monuments, geography and nationalism', *Environment and Planning D: Society and Space* 13: 51–65.

Johnstone, H. (1997) 'Entering the twilight zone', *Asian Business*, 33, 2: 48–51.

Jomo K. S. (1989) *Considerations for a New National Development Strategy*, Kuala Lumpur: Universiti Malaya Institute of Advanced Studies.

—— (1994) 'A nationalist corporative alternative for Malaysian development: lessons from Singapore and other NIC experiences'. Paper presented at *Fourth Malaysia-Singapore Forum*, Universiti Malaya, Faculty of Arts and Social Sciences, 8–11 December.

—— (1995) 'Introduction', in K. S. Jomo (ed.) *Privatizing Malaysia: Rents, Rhetoric, Realities*, Oxford: Westview Press.

—— (2001) 'From currency crisis to recession', in K. S. Jomo (ed.) *Malaysian Eclipse: Economic Crisis and Recovery*, London: Zed Books.

Jomo, K. S. and Ahmad Shabery Cheek (1992) 'Malaysia's Islamic Movements', in J. S. Kahn and F. Loh (eds) *Fragmented Vision: Culture and Politics in Contemporary Malaysia*, Sydney: Allen and Unwin.

Jury, L. (1996) 'Sir Norman Foster's £400m dream – to build Europe's tallest building on a City bomb site', *The Independent*, 10 September: 5.

Kahn, J. S. (1992) 'Class, ethnicity and diversity: some remarks on Malay culture in Malaysia', in J. S. Kahn and F. Loh (eds) *Fragmented Vision: Culture and Politics in Contemporary Malaysia*, Sydney: Allen and Unwin.

—— (1994) 'Subalternity and the construction of Malay identity', in A. Gomes (ed.) *Modernity and Identity: Asian Illustrations*, Victoria: La Trobe University.

—— (1996a) 'Growth, economic transformation, culture and the middle classes in Malaysia', in R. Robison and S. Goodman (eds) *The New Rich in Asia: Mobile Phones, McDonalds and Middle Class Revolution*, London: Routledge.

—— (1996b) 'The middle class as a field of ethnological study', in Muhammad Ikmal Said and Zahid Emby (eds) *Malaysia: Critical Perspectives*, Petaling Jaya: Persatuan Sains Sosial Malaysia.

—— (1997) 'Malaysian modern or anti-anti Asian values', *Thesis Eleven* 50: 15–33.

—— (1998) 'Class, culture and Malaysian modernity', in J. Schmidt, J. Hersh and N. Fold (eds) *Social Change in Southeast Asia*, Harlow: Longman.

Karim Raslan (1996) *Ceritalah: Malaysia in Transition*, Singapore: Times Books.

Karim, W. J. (1995) 'Malaysia's indigenous minorities: discrepancies between nation-building and ethnic consciousness', in Razha Rashid (ed.) *Indigenous Minorities of Peninsula Malaysia: Selected Issues and Ethnographies*, Kuala Lumpur: Intersocietal & Scientific.

Katiman Rostam (1997) 'Industrial expansion, employment changes and urbanization in the peri-urban areas of Klang-Langat Valley, Malaysia', *Asian Profile* 25, 4: 303–15.

Kaur, H. (1996) 'Chicago still sore over tall decision', *Business Times*, 18 April: 2.

—— (1997a) 'PM: MSC designed to meet new needs', *Business Times*, 21 May: 1.

—— (1997b) 'MSC will pave way into Digital Age: Dr M', *Business Times*, 29 March: 26.

Kaur, M. (1997) '30 sen to walk in comfort', *The Star*, 8 May (Metro section): 1.

Kee, H. C. (1993) 'Kuala Lumpur's new city centre', *Wings of Gold*, May: 54–7.

Kelly, P. F. (1999) 'The geographies and politics of globalisation', *Progress in Human Geography* 23: 379–400.

—— (2000) *Landscapes of Globalisation: Human Geographies of Economic Change in the Philippines*, London: Routledge.

—— (2001) 'Metaphors of meltdown: political representations of economic space in the Asian financial crisis', *Environment and Planning D: Society and Space* 19: 719–42.

Kershaw, R. (1982) 'Anglo-Malaysian relations: old roles versus new rules', *International Affairs* 59, 4: 629–48.

Kessler, C. (1999) 'A Malay diaspora? Another side of Dr Mahathir's Jewish problem', *Patterns of Prejudice* 33, 1: 23–42.

Khasnor, J. (1984) *The Emergence of the Modern Malay Administrative Elite*, Kuala Lumpur: Oxford University Press.

Khoo, B. T. (1995) *Paradoxes of Mahathirism: An Intellectual Biography of Mahathir Mohamad*, Kuala Lumpur: Oxford University Press.

—— (2000) 'Unfinished crises: Malaysian politics in 1999', in D. Singh (ed.) *Southeast Asian Affairs 2000*, Singapore: Institute of Southeast Asian Studies.

—— (2002) 'Nationalism, capitalism and "Asian values"', in F. Loh and B. T. Khoo (eds) *Democracy in Malaysia: Discourses and Practices*, Richmond: Curzon.

Khoo, K. J. (1992) 'The grand vision: Mahathir and modernisation', in J. Kahn and F. Loh (eds) *Fragmented Vision: Culture and Politics in Contemporary Malaysia*, Sydney: Allen and Unwin.

King, A. D. (1996) 'Worlds in the city: Manhattan transfer and the ascendance of spectacular space', *Planning Perspectives* 11: 97–114.

Kong, L. and Yeoh, B. S. A. (1996) 'Social constructions of nature in urban Singapore', *Southeast Asian Studies* 34, 2: 402–23.

Krishnamoorthy, M. and Surin, J. A. (1997) 'Online chat', *The Star*, 16 April: 3.

Kua, K. S. (2002) *Malaysian Critical Issues*, Petaling Jaya: Strategic Information Research Development.

Kuala Lumpur City Centre (Holdings) Sdn Bhd (1995) *Meeting Future Needs*, Kuala Lumpur: KLCC (Holdings) Sdn Bhd.

—— (1996a) *Kuala Lumpur City Centre: Business and Investment Opportunities*, Kuala Lumpur: KLCC Marketing Department.

—— (1996b) *Kuala Lumpur City Centre*, Kuala Lumpur: KLCC (Holdings) Sdn Bhd.

Kuala Lumpur International Airport Bhd (1994a) *Lift-off to the Future*, Promotional video produced by Filmwest Sdn Bhd.

—— (1994b) *The Future is Here*, Marketing brochure.

—— (1996) *Kuala Lumpur International Airport Berhad*, Marketing brochure.

—— (1997) *K.L. International Airport: Principal Data*.

Kultermann, U. (1987) 'Architecture in Southeast Asia 4: Malaysia', *Mimar* 26 December: 64–72.

Kusno, A. (2000) *Behind the Postcolonial: Architecture, Urban Space and Political Cultures in Indonesia*, London and New York: Routledge.

—— (2002) 'Architecture After Nationalism: Political Imaginings of Southeast Asian Architects', in T. Bunnell, L. B. W. Drummond and K. C. Ho (eds) *Critical Reflections on Cities in Southeast Asia*, Singapore: Times Academic Press.

Lat (1996) *Lat Gets Lost*, Kuala Lumpur: Berita Publishing.

Latour, B. (1993) *We Have Never Been Modern*, trans. C. Porter, Hemel Hempstead: Harvester Wheatsheaf.

Lau, A. (1991) *The Malayan Union Controversy 1942–48*, Singapore: Oxford University Press.

Lee B. T. (1970) 'Malay urbanization and the ethnic profile of urban areas in Peninsula Malaysia', *Southeast Asian Studies* 8, 2: 224–34.

—— (1987) 'New towns in Malaysia: development and planning policies', in D. R. Phillips and A. G. Yeoh (eds) *New Towns in East and South-east Asia: Planning and Development*, New York: Oxford University Press.

—— (1996) 'Emerging urban trends and the globalizing economy in Malaysia', in F. Lo and Y-M. Yeung (eds) *Emerging World Cities in East Asia*, Tokyo: United Nations University.

Lee, J. (1995) 'PM to launch RM 20bn Putrajaya next week', *Business Times*, 25 August: 4.

Lee, K. Y. (2000) *From Third World to First: The Singapore Story, 1965–2000*, New York: HarperCollins.

Lee, R. L. M. (1990) 'The state, religious nationalism and ethnic rationalization in Malaysia', *Ethnic and Racial Studies* 13, 4: 482–503.

Lefebvre, H. (1991) *The Production of Space*, trans. D. Nicholson-Smith, Oxford: Blackwell.

Leinbach, T. R. (1974) 'The spread of transportation and its impact upon the modernization of Malaya, 1887–1911', *Journal of Tropical Geography* 39: 54–62.

Liak T. K. (1996) 'Mahathir's last hurrah?', in *Southeast Asian Affairs 1996*, Singapore: Institute of Southeast Asian Studies.

Lim, H. K. (1978) *The Evolution of the Urban System in Malaya*, Kuala Lumpur: University of Malaya Press.

Lim, J. (1993) 'An alternative housing approach for the future – the kampongminium concept', Draft copy from author.

Lim J. Y. (1987) *The Malay House: Rediscovering Malaysia's Indigenous Shelter System*, Pinang: Institut Masyarakat.

Lim Kit Siang (1997a) *Cyberlaws in Malaysia*, Petaling Jaya: Democratic Action Party.

—— (1997b) 'Taib Mahmud's double absences', *malaysia.net*. Online posting. Available e-mail: sangkancil@malaysia.net (23 October 1997).

Livingstone, D. N. (1992) *The Geographical Tradition*, Oxford: Blackwell.

Lo, F-c. and Marcotullio, P. J. (2000) 'Globalisation and urban transformations in the Asia-Pacific region: a review', *Urban Studies* 37: 77–111.

Loh, F. (2003) 'Towards a new politics of fragmentation and contestation', in F. Loh and J. Saravanamuttu (eds) *New Politics in Malaysia*, Singapore: Institute of Southeast Asian Studies.

Loh, F. and Kahn, J. S. (1992) 'Introduction: fragmented vision', in J. S. Kahn and F. Loh (eds) *Fragmented Vision: Culture and Politics in Contemporary Malaysia*, Sydney: Allen and Unwin.

Loh, F. and Saravanamuttu, J. (eds) (2003) *New Politics in Malaysia*, Singapore: Institute of Southeast Asian Studies.

Lye, T. P. (2002) 'Forest people, conservation boundaries, and the problem of "modernity" in Malaysia', in G. Benjamin and C. Chou (eds) *Tribal Communities in the Malay World: Historical, Cultural and Social Perspectives*, Singapore: Institute of Southeast Asian Studies.

Lyon, D. (1994) *The Electronic Eye: The Rise of Surveillance Society*, Minneapolis: University of Minnesota Press.

McGee, T. G. (1963) 'The cultural role of cities: A case study of Kuala Lumpur', *Journal of Tropical Geography* 17: 178–96.

—— (1967) *The Southeast Asian City*, London: G. Bell and Sons.

—— (2002) 'Reconstructing "The Southeast Asian City" in an era of volatile globalization', in T. Bunnell, L. B. Drummond and K. C. Ho (eds) *Critical Reflections on Cities in Southeast Asia*, Singapore: Times Academic Press.

McGee, T. G. and McTaggart, W. D. (1967) *Petaling Jaya: A Socio-economic Survey of a New Town in Selangor, Malaysia*, Wellington: Pacific Viewpoint Monograph.

MacInnes, J. (1987) *Thatcherism at Work*, Milton Keynes: Oxford University Press.

McNay, L. (1994) *Foucault: A Critical Introduction*, Cambridge: Polity Press.

Mahathir Mohamad (1970) *The Malay Dilemma*, Singapore: Times Books.

—— (1985) 'Whither Malaysia?', in A. Armour (ed.) *Asia and Japan: The Search For Modernisation and Identity*, London: Athlone Press.

—— (1986) *The Challenge*, Petaling Jaya: Pelanduk Publications.

—— (1992) Speech at the unveiling of the Kuala Lumpur City Centre Project, Selangor Turf Club, 15 September. Online. Available HTTP: http://www.jpm.my/about2.htm (14 January 1996).

—— (1993) 'Malaysia: the way forward', in A. S. A. Hamid (ed.) *Malaysia's Vision 2020: Understanding the Concept, Implications and Challenges*, Petaling Jaya: Pelanduk Publications.

—— (1996a) 'Opening Address'. Speech at the opening of Multimedia Asia on Multimedia Super Corridor (MSC), 1 August, *Multimedia Development Corporation*. Online. Available HTTP: http://www.mdc.com.my/news/archives/index.html (4 November 2000).

—— (1996b) 'Market economy and moral and cultural values: a Malaysian perspective'. Speech at St. Catherine's college, Cambridge, 16 April, *Prime Minister's Office of Malaysia*. Online. Available HTTP: http://www.jpm.my/about2.htm (18 April 1998).

—— (1997a) 'Redeeming dignity of religion' (speech at UMNO General Assembly, Pusat Dagangan Dunia Putra, Kuala Lumpur, 5 September), *Prime Minister's*

Office of Malaysia. Online. Available HTTP: http://www.smpke.jpm.my/prime%20minister/pm-main.htm (18 April 1998).

—— (1997b) Speech at the Los Angeles Conference for Investors on Multimedia Super Corridor, 14 January, UCLA, Prime Minister's Office of Malaysia. Online. Available HTTP: http://www.smpke.jpm.my/about2.htm (18 April 1998).

—— (1998a) *Mahathir Mohamad on the Multimedia Super Corridor*, Subang Jaya: Pelanduk Publications.

—— (1998b) Speech at seminar organised by Mainichi Shimbun, Imperial Hotel, Tokyo, Japan, 20 October, Prime Ministers' Office of Malaysia. Online. Available HTTP: http://www.smpke.jpm.my/pm3.htm (4 November 2000).

Mahathir, M. and Ishihara, S. (1995) *'No' to ieru Aija (The Voice of Asia)*, trans. F. Baldwin, Tokyo: Kodansha.

Malay Mail (1997a) 'You may be next: authorities urged to act against high-rise littering', 30 May: 1.

—— (1997b) 'Binocular watch on tenants', 30 May: 2.

—— (1997c) 'Educate high-rise tenants call', 31 May: 2.

Malaysia (1971) *Second Malaysia Plan, 1971–75*, Kuala Lumpur: Government Printers.

—— (1976) *Third Malaysia Plan, 1976–80*, Kuala Lumpur: Government Printers.

—— (1984) *Mid-Term Review of the Fourth Malaysia Plan, 1981–85*, Kuala Lumpur: Government Printers.

—— (1986) *Fifth Malaysia Plan, 1986–90*, Kuala Lumpur: Government Printers.

—— (1994) *Aboriginal Peoples Act 1954*, Kuala Lumpur: Government Printers.

—— (1996) *Seventh Malaysia Plan, 1996–2000*, Kuala Lumpur: Government Printers.

Malaysian Business (1991) 'The enigmatic "AK"', 1–15 July: 15.

Malhotra, K. (2002) 'Development enabler or disabler? The role of the state in Southeast Asia', in C. J. W.-L. Wee (ed.) *Local Cultures and the 'New Asia': The State, Culture and Capitalism in Southeast Asia*, Singapore: Institute of Southeast Asian Studies.

Markus, T. A. (1985) *Visions of Perfection: The Influence of Utopian Thought Upon Architecture from the Middle Ages to the Present Day*, Glasgow: Third Eye Centre.

—— (1993) *Buildings and Power: Freedom and Control in the Origin of Modern Building Types*, London: Routledge.

Marshall, R. (1998) 'Kuala Lumpur: Competition and the quest for world city status', *Built Environment* 24, 2: 271–84.

Massey, D. (1993) 'Power-Geometry and a progressive sense of place', in J. Bird, B. Curtis, T. Putnam, G. Robertson and L. Tickner (eds) *Mapping The Futures: Local Cultures, Global Change*, New York: Routledge.

Massey, D., Quintas, P. and Wield, D. (1992) *High-Tech Fantasies: Science Parks in Society, Science and Space*, London: Routledge.

Matless, D. (1994) 'Moral geography in Broadland', *Ecumene* 1: 127–55.

—— (1995) 'The art of right living: landscape and citizenship', in S. Pile and N. Thrift (eds) *Mapping the Subject: Geographies of Cultural Transformation*, London: Routledge.

—— (1996) 'New material? Work in cultural and social geography', *Progress in Human Geography* 20, 3: 379–91.

—— (1998) *Landscape and Englishness*, London: Reaktion.

—— (2000) 'Five objects, geographical subjects', in I. Cook, D. Crouch, S. Naylor and J. R. Ryan (eds) *Cultural Turns/Geographical Turns*, Harlow: Pearson Education Limited.

Maznah Mohamad (2000) 'A tale of two cities: monuments to misplaced pride', *Aliran Online*. Online. Available HTTP: http://www.malaysia.net/aliran/high9912e.htm (14 December 2000).

—— (2001) 'The unravelling of a "Malay consensus"', in D. Singh and A. Smith (eds) *Southeast Asian Affairs 2001*, Singapore: Institute of Southeast Asian Studies.

Means, G. (1991) *Malaysian Politics: The Second Generation*, Singapore: Oxford University Press.

MEC (1997) 'Malaysia's greatest achievements', MEC advertisement, 31 August: 40.

Mee, W. (2002) 'Malaysia's multimedia technopole: a nationalist response to globalization and postindustrialism', in T. Bunnell, L. B. Drummond and K. C. Ho (eds) *Critical Reflections on Cities in Southeast Asia*, Singapore: Times Academic Press.

Mellor, W. (2001) 'Field of dreams', *Asiaweek.com*, 17 August. Online. Available HTTP: http://www.asiaweek.com/asiaweek/technology/article/0,8707,170516,00.html (27 August 2001).

Mercer, C. (1997) 'Geographies for the present: Patrick Geddes, urban planning and the human sciences', *Economy and Society* 26, 2: 211–32.

Miller, P. and Rose, N. (1990) 'Governing economic life', *Economy and Society* 19, 1: 1–31.

Milne, R. S. and Mauzy, D. K. (1986) *Malaysia: Tradition, Modernity and Islam*, Boulder: Westview Press.

Milner, A. (1994) *The Invention of Politics in Colonial Malaya: Contesting Nationalism and the Expansion of the Public Sphere*, Cambridge: Cambridge University Press.

Ministry of Education (1997) 'MSC: Catalyst of Information Industry'. Exhibition at National Library, Kuala Lumpur, 8–30 October.

Ministry of Finance Malaysia (2000) '2001 Budget Speech', delivered by Tun Daim Zainuddin, 27 October. Online. Available HTTP: http://www.treasury.gov.my/englishversionbaru/index.htm (14 March 2003).

—— (2002) 'Knowledge-based Economy Masterplan', Kuala Lumpur: Institute of Strategic and International Studies (ISIS). Online. Available HTTP: http://www.treasury.gov.my/englishversionbaru/index.htm (14 March 2003).

Ministry of the Interior (1961) *Statement of Policy Regarding the Administration of the Aborigine Peoples of the Federation of Malaya*, Kuala Lumpur: Federation of Malaya.

Mintz, S. (1985) *Sweetness and Power: The Place of Sugar in Modern History*, New York: Viking Penguin.

Mitchell, D. (2000) *Cultural Geography: A Critical Introduction*, Oxford: Blackwell.

—— (2001) 'The lure of the local: landscape studies at the end of a troubled century', *Progress in Human Geography* 25, 2: 269–81.

Mitchell, K. (1997) 'Transnational discourse: bringing the geography back in', *Antipode* 29: 101–14.

Mitchell, T. (2000) *Questions of Modernity*, Minneapolis: University of Minnesota Press.

Mitchell, W. J. T. (ed.) (1994) *Landscape and Power*, Chicago: University of Chicago Press.

Mitton, R. (1997) 'California Dreamin': PM Mahathir tries to sell a megaproject idea', *Asiaweek*, 7 February: 25.

MOESA (1997) Cartoon on Prangin Mall, *New Straits Times*, 21 March: 12.

Mohamed Arif Nun (1996) 'Putrajaya the Intelligent City: A Strategy Towards the Digital Economy', *MIMOS IT Papers* no. 12.

Mohd Razali Agus (1993) 'Squatters and urban development in Malaysia', *Sarjana (Journal of the Faculty of Arts and Social Science, Universiti Malaya)* 5, 1: 113–42.

Mokhtar, L. I. (1993) 'Urban housing with special emphasis on the squatter population of Kuala Lumpur', in K. Othman (ed.) *Meeting Housing Needs: Issues and Policy Directions*, Kuala Lumpur: Institute of Strategic and International Studies.

Monnet, J. (2001) 'The everyday imagery of space in Los Angeles', in C. G. Salas and M. S. Roth (eds) *Looking for Los Angeles: Architecture, Film, Photography and the Urban Landscape*, Los Angeles: Getty Publications.

Moore, H. L. (1996) 'The changing nature of anthropological knowledge: an introduction', in H. L. Moore (ed.) *The Future of Anthropological Knowledge*, London: Routledge.

Moore, R. (1994) 'The tallest – and guess where it is?', *The Daily Telegraph*, 6 May: 21.

—— (1995) 'The lengths that some countries will go to', *The Daily Telegraph* 20 May: A5.

Morais, V. (1982) *Mahathir: A Profile in Courage*, Singapore: Eastern Universities Press.

Morshidi Sirat (2000) 'Globalising Kuala Lumpur and the strategic role of the producer services sector', *Urban Studies* 12: 2217–240.

Morshidi Sirat and Suriati Ghazali (1999) *Globalisation of Economic Activity and Third World Cities: A Case Study of Kuala Lumpur*, Kuala Lumpur: Utusan.

MSC.Comm (1999a) 'KLCC's hidden architecture', 2, 1: 20–1.

—— (1999b) 'D.I.Y. Travel', 2, 1: 7.

Muhammad Ikmal Said (1992) 'Ethnic perspectives of the Left in Malaysia', in J. S. Kahn and F. Loh (eds) *Fragmented Vision: Culture and Politics in Contemporary Malaysia*, Sydney: Allen and Unwin.

—— (1996) 'Malay nationalism and national identity', in Muhammad Ikmal Said and Zahid Emby (eds) *Malaysia: Critical Perspectives*, Petaling Jaya: Persatuan Sains Sosial Malaysia.

Mulgan, G. (1989) 'The changing shape of the city', in S. Hall and M. Jacques (eds) *New Times: The Changing Face of Politics in the 1990s*, London: Lawrence and Wishart.

Multimedia Development Corporation (1996a) *Investing in Malaysia's MSC: Policies, Incentives and Facilities*, Kuala Lumpur: Multimedia Development Corporation.

—— (1996b) *7 Flagship Applications*, Kuala Lumpur: Multimedia Development Corporation.

—— (1996c) *An Invitation to Malaysia's Multimedia Super Corridor*, Video presentation developed by Neuronet (M) Sdn Bhd.

—— (1997a) *Building the Malaysian Multimedia Super Corridor for World-class Companies*, Kuala Lumpur: Multimedia Development Corporation.

—— (1997b) *Cyberjaya: The Model Intelligent City in the Making*, Kuala Lumpur: Multimedia Development Corporation.

—— (1997c) *An Invitation to Malaysia's MSC: Leading Asia's Information Age*, Kuala Lumpur: Multimedia Development Corporation.

—— (1999) Multimedia Development Corporation website. Online. Available HTTP: http://www.mdc.com.my/ (12 November 2000).

—— (2003) 'IAP History: Year 1997'. Online. Available HTTP: http://www.msc.com.my/mdc/iap/iap1997.asp (7 June 2003).

Murphey, R. (1957) 'New capitals of Asia', *Economic Development and Cultural Change* 12, 3: 216–43.

Murugasu, S. (1999) 'The "leaning" tower episode', *The Star*, 31 August (Section 2): 3.

Muslimedia Webzine (1997) 'Mahathir invites Hollywood to Malaysia', 16–28 February. Online. Available HTTP: http:www.muslimedia.com/archives/sea/sea.htm (23 September 1997).

Mustafa Anuar (2003) 'The role of Malaysia's mainstream press in the 1999 general election', in F. Loh, and J. Saravanamuttu (eds) *New Politics in Malaysia*, Singapore: Institute of Southeast Asian Studies.

Mustafa Anuar and Subramaniam, S. P. (1995) 'The "Good Life"?', *Aliran Monthly* 15, 3: 3–8.

Naidu, G. (1995) 'Infrastructure', in K. S. Jomo (ed.) *Privatizing Malaysia: Rents, Rhetoric, Realities*, Boulder Co: Westview Press.

Naipaul, V. S. (1998) *Beyond Belief: Islamic Excursions Among the Converted Peoples*, London: Little Brown and Company.

Najib, M. A. R. (1997) 'MSC: Catalyst of Information Industry'. Exhibition at National Library, Kuala Lumpur, 8–30 October.

Nash, C. (2000) 'Historical geographies of modernity', in B. Graham and C. Nash (eds) *Modern Historical Geographies*, Harlow: Pearson Education Limited.

Nash, C. and Graham, B. (2000) 'The making of modern historical geographies', in C. Nash and B. Graham (eds) *Modern Historical Geographies*, Harlow: Pearson Education Limited.

Nash, M. (1989) *The Cauldron of Ethnicity in a Modern World*, Chicago: University of Chicago Press.

National Landscape Department (1996) *Landskap Negara: Malaysia Negara Taman*, Kuala Lumpur: Jabatan Landskap Negara.

Nelson, S. (1997) 'The litterbug in you and me', *New Straits Times*, 12 August: 7.

New Perspectives Quarterly (1997) 'From nations to networks: interview with Mahathir Mohamad', 14, 2: 4–8.

New Straits Times (1991) 'Plan for new $7bil city centre', 12 July (City Beat section): 4.

—— (1995) 'KLCC developer urged to employ fresh graduates', 5 May: 2.

—— (1996a) 'Dust problems due to KLCC project', 5 July (Life and Times section): 22.

—— (1996b) 'Lack of community spirit in urban areas', 27 September (Life and Times section): 8.

—— (1997a) '65,000 low-cost houses needed to "clean up" city', 29 July: 8.

—— (1997b) 'Key Yell: Kay Hell: love it, hate it', 1 February (Life and Times section): 1 and 6.

New Sunday Times (1997a) 'KL's strong strides forward', 2 February: 10.

—— (1997b) 'Simple but colourful launch for Cyberjaya', 18 May: 2.

—— (1997c) 'Samy Vellu: Be ready for MSC', 3 August: 3.

—— (1999) 'Shaped by talented young experts', 4 April: 15.

Ng, P. S. C. (1997) 'A job for the locals', *The Sun*, 6 August: 1.

Ngiam, D. (1997) 'Master plan of Shah Alam's Sumurcity ready', *New Straits Times, Business*, 29 June: 27.

Nicholas, C. (1989) 'Theories of development and the underdevelopment of the Orang Asli', *Akademika* 35: 55–68.

—— (1991) 'Orang Asli and development: chased away for a runway', *Pernloi Gah*, October: 5–6.

—— (1993) 'Heeding indigenous rights', *Pernloi Gah*, March: 1–3.

—— (1996) 'A common struggle: Regaining control', in C. Nicholas and R. Singh (eds) *Indigenous Peoples of Asia: Many Peoples, One Struggle*, Bangkok: Asia Indigenous Peoples Pact.

—— (2000) *The Orang Asli and the Contest for Resources: Indigenous Politics, Development and Identity in Peninsula Malaysia*, Copenhagen: International Work Group for Indigenous Affairs.

—— (2002) 'Organising Orang Asli identity', in G. Benjamin and C. Chou (eds) *Tribal Communities in the Malay World: Historical, Cultural and Social Perspectives*, Singapore: Institute of Southeast Asian Studies.

Nonini, D. M. (1997) 'Shifting identities, positioned imaginaries, transnational traversals and reversals by Malaysian Chinese', in A. Ong and D. M. Nonini (eds) *Ungrounded Empires: The Cultural Politics of Modern Chinese Transnationalism*, London: Routledge.

Noordin Sopiee (1996) 'This recently backward region has jumped to the fore', *International Herald Tribune*, 1 March: 6.

Norsaidatual A. M., Harnevie, M. and Valida, A. C. (1999) *Multimedia Super Corridor: A Journey to Excellence in Institutions of Higher Learning*, London: Asean Academic Press.

Ó Tuathail, G. (1997) 'Emerging markets and other simulations: Mexico, the Chiapas revolt and the geofinancial panopticon', *Ecumene* 4: 300–17.

Ó Tuathail, G., Herod, A. and Roberts, S M. (1998) 'Negotiating unruly problematics', in A. Herod, G. Ó Tuathail and S. M. Roberts (eds) *An Unruly World? Globalization, Governance and Geography*, London: Routledge.

Ogborn, M. (1998) *Spaces of Modernity: London's Geographies, 1680–1780*, London: Guilford Press.

Oh, E. (1996) 'Stop work order on Twin Tower if . . .', *Malay Mail*, 9 January: 7.

Ohmae, K. (1992) *Borderless World: Power and Strategy in the Global Marketplace*, London: HarperCollins.

—— (1995) *The End of the Nation State: The Rise of Regional Economies*, London: HarperCollins.

—— (2001) 'How to invite prosperity from the global economy into a region', in A. J. Scott (ed.) *Global City-regions: Trends, Theory, Policy*, New York: Oxford University Press.

Olds, K. (1995) 'Globalization and the production of new urban spaces: Pacific Rim megaprojects in the late 20th Century', *Environment and Planning A*, 27: 1713–43.

—— (2001) *Globalization and Urban Change: Capital, Culture and Pacific Rim Mega-projects*, Oxford: Oxford University Press.

Ong, A. (1987) *Spirits of Resistance and Capitalist Discipline*, Albany: Suny Press.

—— (1996) 'Anthropology, China and modernities: the geopolitics of cultural knowledge', in H. Moore (ed.) *The Future of Anthropological Knowledge*, London: Routledge.

—— (1997) 'Chinese modernities: narratives of nation and of capitalism', in A. Ong and D. M. Nonini (eds) *Ungrounded Empires: The Cultural Politics of Modern Chinese Transnationalism*, London: Routledge.

—— (1999) *Flexible Citizenship: The Cultural Logics of Transnationality*, Durham and London: Duke University Press.

Ong, M. (1990) 'Malaysia, communalism and the political system', *Pacific Viewpoint* 31, 2: 73–95.

Ong-Giger, K. (1997) 'Malaysia's drive into high-technology industries: cruising into the Multimedia Super Corridor?', in D. Singh (ed.) *Southeast Asian Affairs 1997*, Singapore: Institute of Southeast Asian Studies.

Osman, A. (1995) 'Kedah: From rice-bowl state to high-tech industries', in K. P. Gan, K. W. Toh and A. Mathe (eds) *Malaysia Incorporated: Emerging Asian Economic Powerhouse*, Kuala Lumpur: Limkokwing Integrated.

Padman, P. and Lim, A. L. (1989) 'Groups back City Hall's stand on racecourse', *New Straits Times*, 21 August: 8–9.

Parmer, J. N. (1960) *Colonial Labour Policy and Administration: A History of Labour in the Rubber Plantation Industry in Malaya*, Ann Arbor: Monograph of Association of Asian Studies.

Pereira, B. (1997) 'Malaysia rolls out welcome mat to lure top talents home', *The Straits Times* (Singapore), 19 October: 11.

Peter, D. (1997) 'Designer behind KL International Airport', *The Star*, 11 September (Metro section): 2.

Phang S. N., Kuppusamy, S. and Norris, M. W. (1996) 'Metropolitan management of Kuala Lumpur', in J. Ruland (ed.) *The Dynamics of Metropolitan Management in Southeast Asia*, Singapore: Institute of Strategic and International Studies.

Philo, C. and Kearns, G. (1993) 'Culture, history, capital: an introduction to the selling of places', in G. Kearns and C. Philo (eds) *Selling Places: The City as Cultural Capital, Past and Present*, Oxford: Pergamon Press.

Pillai, M. G. G. (1997) 'Is the API being tampered with?', *malaysia.net*. Online posting. Available e-mail: sangkancil@malaysia.net (20 September 1997).

Pillai, P. (1992) *People on the Move: An Overview of Recent Immigration and Emigration in Malaysia*, Kuala Lumpur: Institute of Strategic and International Studies.

Pirie, P. (1976) 'Squatter settlements in Kuala Lumpur', in *Setinggan*, Bangi: Jabatan Antropologi & Sociology, Universiti Kebangsaan Malaysia.

Pred, A. (1995) *ReCognising European Modernities: A Montage of the Present*, London: Routledge.

Pred, A. and Watts, M. (1992) *Reworking Modernity: Capitalisms and Symbolic Discontent*, New Brunswick: Rutgers University Press.

Progressive Architecture (1995) 'Asia bound', April: 44–88.

Putrajaya Corporation (1997) *Putrajaya: The Federal Government Administrative Centre*, Kuala Lumpur: Putrajaya Corporation.

Putrajaya Holdings (1997a) *Putrajaya: An Intelligent Investment*, Kuala Lumpur: Putrajaya Holdings Sdn Bhd.

—— (1997b) 'Birth of a city, birth of a nation', *Sunday Star*, 31 August (special feature advertisement): 23–6.

—— (1997c) 'Government precinct: The federal government's new administrative centre pioneers the era of an electronic government', *The Star*, 28 July (advertisement): 13.

—— (1997d) 'Thankyou', *New Straits Times*, 9 October (advertisement): 5.

Rajah, D. and Perumal, E. (1994) 'Perang Besar is KL's twin city', *The Star*, 3 June (Nation section): 3.

Rajaram, P. K. and Grundy-Warr, C. (2003) 'The Forced Migrant as Homo Sacer: Fear, Brutality and the Withdrawal of Human Rights in the Detention of Irregular Migrants in Malaysia, Thailand and Australia'. Paper presented at the *8th IASFM Conference*, Chiang Mai Thailand, January.

Rajendra, C. (1999) *Shrapnel, Silence and Sand ...*, London: Bogle-L'Ouverture Press.

Ramachandran, S. (1994) *Indian Plantation Labour in Malaysia*, Kuala Lumpur: Institute of Social Analysis.

Ramasamy, P. (1994) *Plantation Labour, Unions, Capital and the State in Peninsula Malaysia*, Kuala Lumpur: Oxford University Press.

—— (1997) 'Villains or victims?', *Aliran Monthly* 17, 2: 7–8.

Ramlan, S. and Singh, S. (1997) 'States get order on Islamic rulings', *New Straits Times*, 9 August: 1.

Ramlan, S., Sariffuddin, T. and Leong, S.-l. (1997) 'Big leap into the future', *New Straits Times*, 18 May: 1.

Ramli Bunting (1991) *Profile: Kampung Orang Asli Busut Salak, Sepang*, Sepang: Jawatan Kuasa Lapangan Terbang Antarabangsa ('International Airport committee').

Rasiah, R. (1995) 'Labour and industrialization in Malaysia', *Journal of Contemporary Asia* 25: 73–92.

Razack, S. H. (2002) 'When place becomes race', in S. H. Razack (ed.) *Race, Space and the Law: Unmapping a White Settler Society*, Toronto: Between the Lines.

Razha Rashid (1995) 'Introduction', in Razha Rashid (ed.) *Indigenous Minorities of Peninsula Malaysia: Selected Issues and Ethnographies*, Kuala Lumpur: Intersocietal & Scientific.

Real Estate Review (1993) 'Country Living: A new trend?', 15: 21–6.

Rehman Rashid (1993) *A Malaysian Journey*, Petaling Jaya: Hikayat.

Reid, A. (1992) 'Economic and social change c. 1400–1800', in *The Cambridge History of Southeast Asia* (Volume 1, Part 2), Cambridge: Cambridge University Press.

Reuters (1997a) 'Malaysia: Mahathir's attack on Jews criticised', malaysia.net. Online posting. Available e-mail: sangkancil@malaysia.net (10 October 1997).

—— (1997b) 'U.S. blasts Malaysian leader's comments on Jews', malaysia.net. Online posting. Available e-mail: sangkancil@malaysia.net (17 October 1997).

Robins, K. (1999) 'Foreclosing on the city? The bad idea of virtual urbanism', in J. Downey and J. McGuigan (eds) *Tecnocities*, London: Sage.

Rose, N. (1996a) 'The death of the social? Re-figuring the territory of government', *Economy and Society* 25: 327–56.

—— (1996b) 'Identity, genealogy, history', in S. Hall and P. du Gay (eds) *Questions of Cultural Identity*, London: Sage.

—— (1999) *Powers of Freedom: Reframing Political Thought*, Cambridge: Cambridge University Press.

Rose, N. and Miller, P. (1992) 'Political power beyond the state: problematics of government', *British Journal of Sociology* 43: 172–205.

Ruland, J. (1992) *Urban Development in Southeast Asia: Regional Cities and Local Government*, Boulder CO: Westview Press.

Ruslan Khalid (1989) 'Help turn turf club to a "People's Park"'. Letter to *New Straits Times*, 28 August: 15.

Sabri Zain (2000) *Face Off: A Malaysian Reformasi Diary (1998–99)*, Singapore: BigO Books.

Sagong bin Tasi & Ors v Kerajaan Negeri Selangor (2002) *Malaysian Law Journal*, 2: 591.

Saravanamuttu, J. (1989) 'Kelas menengah dalam politik Malaysia: tonjolan perkauman atau kepentingan kelas', *Kajian Malaysia* 7, 1&2: 106–26.

—— (2003) 'The eve of the 1999 general election: from the NEP to *reformasi*', in F. Loh and J. Saravanamuttu (eds) *New Politics in Malaysia*, Singapore: Institute of Southeast Asian Studies.

Sardar, Z. (1998) 'The sport of Mahathir Mohamed', *New Statesman*, 11 September: 17–18.

—— (2000) *The Consumption of Kuala Lumpur*, London: Reaktion Books.

Sassen, S. (1991) *The Global City: New York, London, Tokyo*, Chichester: Princeton University Press.

Schein, R. (1997) 'The place of landscape: a conceptual framework for interpreting an American scene', *Annals of the Association of American Geographers* 87, 4: 660–80.

Scientific American (1994a) 'MIMOS: Bridging Technology and Business', April (Malaysia Advertising Section): 10.

—— (1994b) 'Technology Park Malaysia: Home for Technology Development', April (Malaysia Advertising Section): 12.

Scott, A. J., Agnew, J., Soja, E. W. and Storper, M. (2001) 'Global city-regions', in A. J. Scott (ed.) *Global City-regions: Trends, Theory, Policy*, New York: Oxford University Press.

Scott, J. C. (1998) *Seeing Like a State: How Certain Schemes to Improve the Human Condition Have Failed*, New Haven and London: Yale University Press.

Seabrook, J. (1996) 'Not a lot to smile about', *New Statesman & Society*, 19 January: 20–1.

Searle, P. (1999) *The Riddle of Malaysian Capitalism: Rent-Seekers or Real Capitalists?*, St. Leonards: Allen & Unwin.

See Yee Ai (1997) 'Courting Malaysian expats', *The Star*, 6 October (Section 2): 4–5.

Shaharuddin, H. I. (1991) 'Instilling spiritual values in Vision 2020', *New Straits Times*, 21 August: 11.

Sham Sani (1989) 'Aspek sosio-ekonomi masyarakat Temuan di Bukit Tampoi', *Akademika* 35: 87–96.

Shamsul A. B. (1986) *From British to Bumiputera Rule: Local Politics and Rural Development in Peninsula Malaysia*, Singapore: Institute of Southeast Asian Studies.

—— (1988) 'Kampung: Antara Kenyataan Dengan Nostalgia', *Dewan Masyarakat*, December: 4–5.

—— (1996a) 'Debating about identity in Malaysia: a discourse analysis', *Southeast Asian Studies* 34, 3: 476–99.

—— (1996b) 'Nations-of-intent in Malaysia', in S. Tonneson and H. Antlov (eds) *Asian Forms of the Nation*, Richmond: Curzon.

—— (1997) 'Ethnicity, class or identity? In search of a new paradigm in Malaysian studies'. Paper presented at *First International Malaysian Studies Conference*, Universiti Malaya, 11–13 August.

Shamsul, B. and Lee, B. T. (1988) *FELDA: 3 Decades of Evolution*, Kuala Lumpur: FELDA.

Shareem Amry (1997a) 'Public apathy about cleanliness symptomatic of slow social development', *New Straits Times*, 12 August: 2.

—— (1997b) 'All eyes on Malaysia's Multimedia Super Corridor', *New Straits Times*, 26 June: 2.

—— (1997c) 'City of future with a strong local imprint', *New Sunday Times*, 13 April: 16.

Short, J. R., Breitback, C., Buckman, S. and Essex, J. (2000) 'From world cities to gateway cities', *City* 4, 3: 317–40.

Sia, A. (1997a) 'Urban Utopia?', *The Star*, 28 July (Section 2): 1–4.

—— (1997b) 'Deepavali's different now', *The Star*, 25 October (Weekender section): 5.

Sidaway, J. D. (2002) *Imagined Regional Communities: Integration and Sovereignty in the Global South*, London: Routledge.

Sidaway, J. D. and Pryke, M. (2000) 'The strange geographies of emerging markets', *Transactions of the Institute of British Geographers* 25: 187–201.

Siegel, L. and Markoff, J. (1985) *The High Cost of High Tech: The Dark Side of the Chip*, New York: Harper and Row.

Sigley, G. (1996) 'Governing Chinese bodies: the significance of studies in the concept of governmentality for the analysis of government in China', *Economy and Society* 25, 4: 457–82.

Singh, K. (1997a) 'Conducive for creativity', *The Edge*, 26 May (City and Country section): 1.

—— (1997b) 'Natural extension of Kelang Valley's urban fabric', *The Edge*, 5 May (City and Country section): 7.

Sioh, M. (1998) 'Authorizing the Malaysian rainforest: configuring space, contesting claims and conquering imaginaries', *Ecumene* 5, 2: 144–66.

Sivarajan, A. (1995) 'Eviction – its effects on the health of the plantation community', in *Proceedings of Estate Health and Vision 2020*. Conference organised by Malaysian Medical Association, Faculty of Medicine, University of Malaya, 26 November.

Sloane, P. (1999) *Islam, Modernity and Entrepreneurship Among the Malays*, Houndmills: Macmillan.

Smith, C., Burke, H. and Ward, G. K. (2000) 'Globalisation and indigenous peoples: Threat or empowerment', in C. Smith and G. K. Ward (eds) *Indigenous Cultures in an Interconnected World*, St. Leonards, NSW: Allen and Unwin.

Smith, M. P. (2001) *Transnational Urbanism: Locating Globalization*, Blackwell, Oxford.

Smith, W. (1999) 'The contribution of a Japanese firm to the cultural construction of the new rich in Malaysia', in M. Pinches (ed.) *Culture and Privilege in Capitalist Asia*, London: Routledge.

Sobri, S. (1985) 'City within a city', *Majallah Akitek*, February: 20–25.

Spaeth, A. (1996) 'Bound for glory', *Time*, 9 December: 4–8.

Star (1997a) 'Golden Hope corporate campus to be a MSC landmark', 6 May (Business section): 14.

—— (1997b) 'Ban currency manipulation, says Mahathir', 8 September: 2.

Stenson, M. (1980) *Class, Race and Colonialism in West Malaysia: The Indian Case*, St. Lucia: University of Queensland Press.

Stoler, A. L. (1995) *Race and the Education of Desire: Foucault's History of Sexuality and the Colonial Order of Things*, Durham and London: Duke University Press.

Sudjic, D. (1996) 'The height of madness', *The Guardian*, 1 March (Review section): 1–3.

Sullivan, P. (1985) 'A critical appraisal of historians of Malaysia: the theory of society implicit in their work', in R. Higgot and R. Robison (eds) *Southeast Asia: Essays in the Political Economy of Structural Change*, London: Routledge.

Sun (1997) 'Nationwide link', 16 April: 1.

Sunday Star (1989) 'Elyas: Park plan for racecourse still on', 20 August: 6.

—— (1997) 'Fitting old values into a new age', 4 May: 17.

Sweeney, J. (1998) 'Migrants poisoned and deported', *The Observer*, 26 April: 1.

Syed Husin Ali (1996) *Two Faces: Detention Without Trial*, Kuala Lumpur: Insan.

Tan, A. A. L. (1996) 'Cultural modernism – new trend in symbiotic architecture', *Building Property Review*, Sept/Oct: 72–5.

Tan, E. (1996) 'Wetland sanctuary to be set up soon', *New Straits Times*, 2 August: 1&7.

Tan, J. (1993) 'A Temuan Odyssey continues: Sepang folk forced to give up their development for others', *Pernloi Gah*, March: 5–8.

—— (1997) 'Definitions blur over fashionable "civil society"', *New Sunday Times*, 22 June: 12.

Tan, K. P. (1994) 'Golden Hope likely to get RM 450m windfall', *The Star*, 4 June: 16.

Tan, S. B. (1992) 'Counterpoints in the performing arts of Malaysia', in J. Kahn and F. Loh (eds) *Fragmented Vision: Culture and Politics in Contemporary Malaysia*, Sydney: Allen and Unwin.

Tay, K. S. (1989) *Megacities in the Tropics: Towards an Architectural Agenda for the Future*, Singapore: Institute of Southeast Asian Studies.

Taylor, P. J. (1996) 'Embedded statism and the social sciences: opening up to new spaces', *Environment and Planning* A 28: 1917–28.

—— (1999) *Modernities: A Geohistorical Interpretation*, Minneapolis: University of Minnesota Press.

Telekom Malaysia and Everest '97 Team (1997) 'Everest is ours', *New Straits Times*, 27 May: 7.

Teoh, S. (1997) 'A question of freedom . . .', *Investors Digest* March: 10–11.

Thongchai, W. (1994) *Siam Mapped: A History of the Geo-body of a Nation*, Hawaii: University of Hawaii Press.

Thrift, N. (1996a) 'New urban eras and old technological fears: reconfiguring the goodwill of electronic things', *Urban Studies* 33, 8: 1463–93.

—— (1996b) *Spatial Formations*, London: Sage.

—— (1997) 'The still point: resistance, expressive embodiment and dance', in S. Pile and N. Thrift (eds) *Geographies of Resistance*, London: Routledge.

—— (2000a) 'Animal spirits: performing cultures in the new economy', *Annals of the Association of American Geographers* 90: 674–92.

—— (2000b) 'Introduction: Dead or alive?', in I. Cook, D. Crouch, S. Naylor and J. R. Ryan (eds) *Cultural Turns/Geographical Turns*, Harlow: Pearson Education Limited.

—— (2001) 'Non-representational theory', in R. J. Johnston, D. Gregory, G. Pratt and M. Watts (eds) (fourth edition) *The Dictionary of Human Geography*, Oxford: Blackwell.

Thrift, N., Driver, F. and Livingstone, D. (1995) 'Editorial: The geography of truth', *Environment and Planning D: Society and Space* 13: 1–3.

Time (1996) 'Master Planner', 9 December: front cover.

Tregonning, K. G. (1966) 'Singapore and Kuala Lumpur: A politico-geographical contrast', *Pacific Viewpoint* 7, 2: 238–41.

Trezzini, B. (2001) 'Embedded state authority and legitimacy: piecing together the Malaysian development puzzle', *Economy and Society* 30, 3: 324–53.

Tsou, P.-C. (1967) *Urban Landscape of Kuala Lumpur: A Geographical Survey*, Singapore, Nanyang University.

Turner, J. (1997) 'Upwardly mobile', *Independent On Sunday*, 23 February: 36.

United Nations Human Settlements Programme (1996) *The Habitat Agenda*. UN-Habitat. Online. Available HTTP: http://www.unhabitat.org/declarations/habitat_agenda.asp (25 May 2003).

Universiti Pertanian Malaysia (1995) *Laporan Penilaian Kesan Alam Sekitar di Daerah Sepang, Selangor Darul Ehsan*, Kuala Lumpur: Government Printers.

Utusan Konsumer (1997) 'What is there to celebrate?', 27, 10: 20.

Vale, L. (1992) *Architecture, Power and National Identity*, New Haven: Yale University Press.

Van Leeuwen, T. (1992) *The Skyward Trend of Thought*, Cambridge MA: MIT Press.

Vidal, J. (1997) 'The thousand mile shroud', *The Guardian*, 8 November (Weekend section): 14–24.

von Einsiedel, N. (1997) 'Ensuring our urban futures: implementing the Habitat 2 agenda'. Paper presented at *South-South Mayor's Conference*, Putra World Trade Centre, Kuala Lumpur, 4 July.

Wade, R. (1990) *Governing the Market: Economic Theory and the Role of Government in East Asian Industrialization*, Princeton: Princeton University Press.

Wang, G. (2001) *Only Connect! Sino-Malay Encounters*, Singapore: Times Academic Press.

Watson, C. W. (1996) 'The construction of the post-colonial subject in Malaysia', in S. Tonnesson and H. Antlov (eds) *Asian Forms of the Nation*, Richmond: Curzon.

Watts, M. (2003) 'Alternative modern – development as cultural geography', in K. Anderson, M. Domosh, S. Pile and N. Thrift (eds) *Handbook of Cultural Geography*, London: Sage.

Webb, B. (1996) 'The less-than-super-model', *New Statesman & Society*, 26 January: 17.

Wee, C. J. W.-L. (1996) 'The "clash" of civilizations? Or an emerging "East Asian modernity"?', *SOJOURN* 11, 2: 211–30.

—— (2002) 'Introduction: Local cultures, economic development and Southeast Asia', in C. J. W.-L. Wee (ed.) *Local Cultures and the 'New Asia': The State, Culture and Capitalism in Southeast Asia*, Singapore: Institute of Southeast Asian Studies.

Wheatley, P. (1961) *The Golden Khersonese: Studies in the Historical Geography of the Malay Peninsula Before A.D. 1500*, Kuala Lumpur: University of Malaya Press.

Williams-Hunt, A. (1995) 'Land conflicts: Orang Asli ancestral laws and state policies', in Razha Rashid (ed.) *Indigenous Minorities of Peninsula Malaysia: Selected Issues and Ethnographies*, Kuala Lumpur: Intersocietal & Scientific.

Willford, A. (2002) 'Weapons of the meek: Ecstatic ritualism and strategic ecumenism among Tamil Hindus in Malaysia', *Identities: Global Studies in Culture and Power* 9: 247–80.

Winner, L. (1992) 'Silicon Valley mystery house', in M. Sorkin (ed.) *Variations on a Theme Park: The New American City and the End of Public Space*, New York: Hill and Wang.

World Bank (1993) *The East Asian Miracle: Economic Growth and Public Policy*, New York: Oxford University Press.

—— (1999) *Global Urban and Local Government Strategy*. Online. Available HTTP: http://www.worldbank.org/html/fpd/urban/strategy/chap1.pdf (9 November 2000).

Yao S. (2001a) 'Introduction', in S. Yao (ed.) *House of Glass: Culture, Modernity and the State in Southeast Asia*, Singapore: Institute of Southeast Asian Studies.

—— (2001b) 'Modernity and Mahathir's rage: theorizing state discourse of mass media in Southeast Asia', in S. Yao (ed.) *House of Glass: Culture, Modernity and the State in Southeast Asia*, Singapore: Institute of Southeast Asian Studies.

Yeang, K. (1987) *Tropical Urban Regionalism*, Singapore: Mimar.

—— (1992) *The Architecture of Malaysia*, Amsterdam: Pepin Press.

—— (1994) *Bioclimatic Skyscrapers*, Berlin: Aedes.

Yeoh, B. S. A. (1996) *Contesting Space: Power Relations and the Urban Built Environment in Colonial Singapore*, Kuala Lumpur: Oxford University Press.

—— (2000) 'Historical geographies of the colonised world', in B. Graham and C. Nash (eds) *Modern Historical Geographies*, Harlow: Pearson Education Limited.

—— (2001) 'Postcolonial cities', *Progress in Human Geography* 25, 3: 456–68.

Yeoh, S. G. (2001) 'Creolized utopias: squatter colonies and the post-colonial city in Malaysia', *Sojourn*, 16: 102–24.

Yeung, H. Y. C. (1998) 'Capital, state and space: contesting the borderless world', *Transactions of the Institute of British Geographers* 23: 291–309.

Yong, D. (1991) 'Move to re-alienate turf club land', *The Star*, 21 March: 3.

Yong, A. and Choong, R. (1997) 'The first 29: Well-known companies among MSC pioneer batch', *The Star*, 25 April: 1.

Yusof Ghani (1997) 'Last of the Temuan', *The Sun*, 6 August (Magazine section): 4–5.

Zahid Emby (1990) 'The Orang Asli regrouping scheme: Converting swiddeners to commercial farmers', in V. T. King and M. Parnwell (eds) *Margins and Minorities: The Peripheral Areas and Peoples of Malaysia*, Hull: Hull University Press.

Zainuddin, M. (1995) 'Urban growth, city development and the new planning doctrine', *Planning Malaysia* 1: 11–22.

—— (1997) 'Planning of Cyberjaya: the multimedia city for Malaysia'. Paper presented at *South-South Mayors Conference: Developing Solutions for the Cities of the 21st Century*, Putra World Trade Centre, Kuala Lumpur, 3–4 July.

Zawawi, I. (1996a) 'The making of subaltern discourse in the Malaysian nation-state: new subjectivities and the poetics of Orang Asli dispossession and identity', *Southeast Asian Studies* 34, 3: 568–600.

—— (ed.) (1996b) *Kami Bukan Anti-pembangunan: Bicara Orang Asli Menuju Wawasan 2020*, Bangi: Persatuan Sains Sosial Malaysia.

Index

Printed and bound by CPI Group (UK) Ltd, Croydon, CR0 4YY

01/11/2024

01782627-0004